# Philosophical Perspectives on the Engineering Approach in Biology

*Philosophical Perspectives on the Engineering Approach in Biology* provides a philosophical examination of what has been called the most powerful metaphor in biology: The machine metaphor. The chapters collected in this volume discuss the idea that living systems can be understood through the lens of engineering methods and machine metaphors from both historical, theoretical, and practical perspectives.

In their contributions the authors examine questions about scientific explanation and methodology, the interrelationship between science and engineering, and the impact that the use of engineering metaphors in science may have for bioethics and science communication, such as the worry that its wide application reinforces public misconceptions of the nature of new biotechnology and biological life. The book also contains an introduction that describes the rise of the machine analogy and the many ways in which it plays a central role in fundamental debates about e.g. design, adaptation, and reductionism in the philosophy of biology.

The book will be useful as a core reading for professionals as well as graduate and undergraduate students in courses of philosophy of science and for life scientists taking courses in philosophy of science and bioethics.

**Sune Holm** is Associate Professor in Philosophy at the Department of Food and Resource Economics at the University of Copenhagen. His research mainly focuses on topics in ethics and philosophy of science relating to biology, biotechnology, and artificial intelligence.

**Maria Serban** is a lecturer in philosophy at University of East Anglia. She is a philosopher of science focusing on modelling practices in the life sciences.

**History and Philosophy of Biology**
Series Editor: Rasmus Grønfeldt Winther is Associate Professor of Philosophy at the University of California, Santa Cruz (UCSC).

This series explores significant developments in the life sciences from historical and philosophical perspectives. Historical episodes include Aristotelian biology, Greek and Islamic biology and medicine, Renaissance biology, natural history, Darwinian evolution, Nineteenth-century physiology and cell theory, Twentieth-century genetics, ecology, and systematics, and the biological theories and practices of non-Western perspectives. Philosophical topics include individuality, reductionism and holism, fitness, levels of selection, mechanism and teleology, and the nature-nurture debates, as well as explanation, confirmation, inference, experiment, scientific practice, and models and theories vis-à-vis the biological sciences.

Authors are also invited to inquire into the "and" of this series. How has, does, and will the history of biology impact philosophical understandings of life? How can philosophy help us analyze the historical contingency of, and structural constraints on, scientific knowledge about biological processes and systems? In probing the interweaving of history and philosophy of biology, scholarly investigation could usefully turn to values, power, and potential future uses and abuses of biological knowledge.

The scientific scope of the series includes evolutionary theory, environmental sciences, genomics, molecular biology, systems biology, biotechnology, biomedicine, race and ethnicity, and sex and gender. These areas of the biological sciences are not silos, and tracking their impact on other sciences such as psychology, economics, and sociology, and the behavioral and human sciences more generally, is also within the purview of this series.

**Biological Identity**
Perspectives from Metaphysics and the Philosophy of Biology
*Edited by Anne Sophie Meincke and John Dupré*

**Philosophical Perspectives on the Engineering Approach in Biology**
Living Machines?
*Edited by Sune Holm and Maria Serban*

For more information about this series, please visit: www.routledge.com/History-and-Philosophy-of-Biology/book-series/HAPB

# Philosophical Perspectives on the Engineering Approach in Biology

Living Machines?

Edited by
**Sune Holm and Maria Serban**

Routledge
Taylor & Francis Group

LONDON AND NEW YORK

First published 2021
by Routledge
2 Park Square, Milton Park, Abingdon, Oxon OX14 4RN

and by Routledge
52 Vanderbilt Avenue, New York, NY 10017

*Routledge is an imprint of the Taylor & Francis Group, an informa business*

*British Library Cataloguing-in-Publication Data*
A catalogue record for this book is available from the British Library

*Library of Congress Cataloging-in-Publication Data*
Names: Holm, Sune, editor.
Title: Philosophical perspectives on the engineering approach in biology : Living machines? / edited by Sune Holm and Maria Serban.
Description: First edition. | Abingdon, Oxon ; New York : Routledge, 2020. | Series: History and philosophy of biology | Includes bibliographical references and index.
Identifiers: LCCN 2020005123 (print) | LCCN 2020005124 (ebook) | ISBN 9780815380788 (hardback) | ISBN 9781351212243 (ebook)
Subjects: LCSH: Biology—Philosophy. | Bioengineering—Moral and ethical aspects. | Bioethics.
Classification: LCC QH331 .P466 2020 (print) | LCC QH331 (ebook) | DDC 570.1—dc23
LC record available at https://lccn.loc.gov/2020005123
LC ebook record available at https://lccn.loc.gov/2020005124

ISBN: 978-0-8153-8078-8 (hbk)
ISBN: 978-0-367-49146-8 (pbk)
ISBN: 978-1-351-21224-3 (ebk)

Typeset in Times New Roman
by Apex CoVantage, LLC

Visit the eResources: http://www.routledge.com/9780815380788

# Contents

# Contributors

**William Bechtel**, Department of Philosophy, University of California, San Diego.

**Brett Calcott**, Honorary Associate, University of Sydney.

**Andreas T. Christiansen**, Section for Philosophy, Department of Communication, University of Copenhagen.

**Sara Green**, Assistant Professor, Section for History and Philosophy of Science, Department of Science Education, University of Copenhagen.

**Sune Holm**, Associate Professor, Department of Food and Resource Economics, University of Copenhagen.

**Louisa Holt**, Research Assistant, Department of Food and Resource Economics, University of Copenhagen.

**Tarja Knuuttila**, Professor of Philosophy of Science, Department of Philosophy, University of Vienna.

**Arnon Levy**, Department of Philosophy, The Hebrew University of Jerusalem.

**Andrea Loettgers**, Department of Philosophy, University of Vienna.

**Daniel J. Nicholson**, Konrad Lorenz Institute for Evolution and Cognition Research; Centre for the Study of Life Sciences (Egenis), University of Exeter.

**Jessica Riskin**, Professor of History, Stanford University.

**Maria Serban**, Postdoc, Institute for Philosophy, Literature, the History of Science and Technology, Technical University Berlin.

# Acknowledgments

Some of the chapters in this book were presented at a workshop on the machine analogy in biology at the University of Copenhagen in 2016. The workshop was organized as part of the project *Living Machines? Prospects and Limitations of the Engineering Approach to the Organism*. We would like to thank Independent Research Fund Denmark grant number DFF – 4180–00146 for funding the project and the workshop. Also, we would like to thank the authors of the chapters in this volume for their encouragement and contributions. Finally, we would like to thank Sofie Bahnsen for her help with the task of making the complete manuscript ready for publication.

# Introduction

*Louisa Holt, Sune Holm, and Maria Serban*

## The rise of the machine analogy[1]

For as long as philosophers and scientists have asked the question, "What is an organism?" there have been those that have answered, "It is a machine." For as long as this answer has been given, it has been disputed, the disparities between organism and machine being at least equal to their similarities. And yet the idea has remained, constantly reinventing itself to reflect both our evolving understanding of organisms and the technology of the day. Discredited in one form, it reemerges in another. Discredited in every form, it shapeshifts from ontological stance to heuristic tool or rhetorical device.

The first known comparison between organism and machine is usually attributed to the Greek physician Hippocrates, who likened the human body to a water clock. The latest technology in 400 BC, water clocks were hydraulic devices whose various parts were controlled by a carefully balanced flow of water in order to tell the time. Reflecting on these technologies, Hippocrates proposed that the human body was made up of four humors that existed as liquids: blood, phlegm, black bile, and yellow bile. And like the liquid in the clock, the four humors had to be kept in balance for the body to function (Jouanna 2012).

In the 14th century, mechanical clocks started to appear. The first were driven by a simple system of weights. These were soon replaced by springs and finally, in the 17th century, by the spectacular, complex, and precise pendulum clocks that would inspire Descartes, La Mettrie, and Newton. It is hard for us now to imagine how truly marvelous this technology must have seemed. Objects that for all time had been inert and completely dependent on a human operator now sprang into motion, asserting some capacity for self-directed movement. As the philosopher Comenius put it, "Is it not a truly marvellous thing that a machine, a soulless thing, can move in such life-like, continuous, and regular manner? Before clocks were invented would not the existence of such things have seemed as impossible as that trees could talk or stones speak? Yet everyone can see that they exist now" (in McReynolds 1980). For the first time, a link was forged between the inanimate and animate worlds. This link paved the way for a new approach to understanding organisms; it was an approach that did not require vital spirits or mysterious gods. Just as we explain how the hands of a clock move by referencing only its inanimate parts, so, too, might we explain how an organism can walk or talk by referencing interactions between the inert matter out of which it is made. In the face of this new technology, the idea that an organism was like a simple water clock quickly seemed outdated. Surely if the organism was a type of machine, it must be like a pendulum clock, belonging to the same family as the spectacular Strasbourg astronomical clock, only ever more complex.

The core of the analogy between organism and clock turns on the idea that organisms can be explained in terms of regular, recurring, automatically controlled events among their parts. It positions the body as a purely mechanical thing, which can only be understood by studying it as such. It is hard to overstate the importance of this idea to much of the scientific progress that has occurred since the 17th century. First, the rise of the mechanical view served a huge blow to vital theories of life that had dominated since Aristotle. But more importantly for the scientist, it justified adopting engineering methods within biology (Brito and Marques 2014). If the organism is a type of machine, then it can be studied like a machine: to understand an organism we need only decompose it into its constituent parts and then identify the functions of those parts. Take the example of medicine. A basic assumption of the field, so pervasive we scarcely notice it, is that we can repair a human body in much the same way as we might repair a broken engine – locate the faulty part and either fix or repair it. The very success of this approach attests to its legitimacy.

Although the machine analogy would continue to evolve as technology itself evolved, all later transformations of the idea would retain this kernel of an idea: that we can understand the whole system by studying its most basic parts and how they interact. What now seemed pertinent was not the claim that the body was a type of machine but identifying which type of machine the body was.

Hot on the heels of the age of mechanics came the age of heat and steam. In the 16th century, the first vacuum and air pumps started to appear. In the 17th century, this technology was harnessed by Denis Papin to build the first steam digester, a machine that confined steam in a tight space until a high pressure was created. In Papin's machine, a combination of heat and water was used to create a vacuum, causing atmospheric pressure to build up against a piston, resulting in motive power. Then, the first steam engines appeared, machines that harnessed that pressure to produce mechanical work (Dickinson 1939). At the same time, and very much hand in hand with these technological advancements, the theory of thermodynamics was rapidly developing. This led to a new version of the machine analogy, first popularized by von Bertalanffy (1968) but also elaborated on by Maturana and Varela in their theory of organisms as "autopoietic machines" (1973/1980). Like the theory of thermodynamics itself, the new analogy was comprised of two parts. The first, based on the first law of thermodynamics, had to do with the conservation of energy. The primary idea was that animals, like machines, must be fed by a constant flow of fuel, which must be burnt, so that the overall energy level remains stable. The second analogy was based on the second law of thermodynamics, the tendency of all systems toward disorder and entropy. In this respect, organisms and machines seem to be very similar. They both seem to act against the general tendency toward disorder. To the contrary, organisms and machines are both highly organized systems.

Meanwhile, in the 18th century, the first telegraphs appeared, systems that involved the transmission of signals, not just energy. This technology led thinkers to the idea that organisms are not just connected to the world by flows of energy but also by flows of information. By the mid-20th century, this new analogy would find firm expression in the developing field of cybernetics. Popularized largely by Wiener, cybernetics was concerned with systems that process and react to environmental information to achieve their goals. As early as 1944, Wiener was organizing meetings that brought together engineers working in communication and control with biologists and medical researchers. In his 1948 manifesto (Wiener 1948), *Cybernetics: Or Control and Communication in the Animal and the Machine*, Wiener offered a theory of organisms that emphasized messages, codes, information, and feedback loops. The new analogy consisted of three parts. First, the machine has sensory organs just like an organism. It has thermometers, light receptors, microphones, and all sorts of other tools used to register various signals. Second, the machine has motor functions just like the organism. These are carried out by whatever device the machine uses to create an output – an engine or heating element, for instance. Finally, between the sensory and motor organs, the machine has a central regulating system that coordinates incoming messages from the outside world and from its internal organs and then directs the system to bring about the appropriate reaction.

Finally, in the 1940s Turing invented the digital computer. If cybernetics told us that studying life involves studying the reception, transmission, storage, transformation, and use of information, computation provided a fleshed-out theory for how that information processing actually works. It did so by appealing to algorithms: finite sets of rules that transform data sets in predetermined ways. It is this appeal that sits behind one of the most prolific versions of the machine analogy in modern science: the claim that DNA is a type of computer program, a genetic code. Building directly on Wiener's conception of the organism, von Neumann presented his vision of a biological computing machine able to reproduce itself by following a set of instructions encoded as a set of genes. The gene was compared to the information tape of a Universal Turing Machine; it was envisioned as encoding a set of instructions for programming an organism (Cobb 2013). It is an idea that predates Watson and Crick but which remains hugely influential to this day. Just six weeks after they first described its double-helix structure, Watson and Crick would write, "The precise sequence of the bases is the code which carries the genetic information" (1953). This gave rise to the great coding challenge of the 1950s and 1960s, trying to ascertain which letters in the DNA code corresponded to which amino acids. It was a challenge that was finally cracked by Nirenberg and Matthaei, along with their collaborators, in 1967. There is no doubt that framing the challenge of DNA transcription in the familiar terminology of a code helped lead to one of the most important scientific discoveries in the last century. This is true because it seems inarguable that DNA really does contain information (Griffiths 2001).

But there is a troubling implication in conceptualizing DNA as a type of code; it seems to suggest that development, like the activity of a computer, is automatic and predetermined, that you can explain the traits and behavior of an organism by reference to its genes alone or, alternatively, that genes have some kind of special causal power. In short, it can quickly lead to genetic determinism, which in turn is linked to the morally problematic claim that socially significant claims we care about are largely the result of the operation of genes and cannot be changed during development by the environment, learning, or other human interventions. Basing our understanding of genetics on the idea that genes are (like) blueprints or programs makes deterministic readings of claims about the role of genes in development almost inevitable. This is not to deny that "information talk" has many legitimate applications in molecular biology and developmental biology. But the mathematical meaning of "information" is too unintuitive and too far from the usual meaning of the word to become part of the public understanding of how genes work. Therefore, the analogy, while powerful in the applications it generates, remains potentially misleading. One practical solution proposed to improve this situation is the popularization of other analogies and ideas from biological research that can act as a counterweight to misinterpretations of the code/information analogy (Oyama et al. 2001).

The machine analogy and the mechanistic approach to biology it inspired have brought with them undeniable success. This success has turned on an important truth that sits at the heart of the machine analogy: both organisms and machines are constituted by functionally differentiated parts, parts that have purpose. Through time, machines have provided biologists with inspiration as to how those parts might be put together. And while no technological prototype serves as the perfect model, each arguably contains a kernel of truth that can be borrowed, applied, bent, and contradicted. All of this, including the criticisms and refutations, can increase our understanding of the nature of organisms. However, the strength of the analogy lies not in its ontological veracity but in the bridge it builds to engineering methods.

Focusing on the functional nature of biological systems has availed biologists with two key methods of scientific discovery, both inspired by engineering practices. In reverse engineering, biologists start by decomposing a physical system into its parts and then investigating their functions and the general design principles that hold the system together. In forward engineering, biologists start by identifying a need of the target system and then go looking for the mechanism that fulfills that function. The contribution these approaches have made to our current understanding of organisms is immense. In this way, when the analogy fails ontologically, its value as an epistemic heuristic shifts into focus. The analogy struggles in one incarnation but resurrects itself in another guise. Despite its flaws, it persists as a central feature of many of the most vibrant debates in modern philosophy of biology. To understand these debates, one must understand the machine analogy itself. In the next section, we will briefly outline some of these debates and the role the machine analogy plays within them.

## The machine analogy in philosophy of biology

### *Design in nature*

The history of machine analogies traced above is a series of responses to the question, "What type of thing is an organism?" Is it a mechanism? A thermodynamic system? A cybernetic system? A computer? But there is another side to the machine analogy, which focuses on how an organism is made rather than what an organism is. Let us call this the design metaphor. It is important to be clear that these are not separate analogies. On the contrary, one of the most fundamental features of a machine is the fact that it has been designed and produced by an intelligent and intentional being. To even suggest that an organism is a type a machine is arguably to suggest that it has been designed or made. In this way, the machine analogy of the organism developed by Descartes and his contemporaries in the 17th century directly paved the way for the argument from design in the centuries to come, no doubt the most famous expression of which is found in Paley's *Natural Theology* (1836). Comparing the parts of a watch, designed to be so perfectly suited to their tasks, with the remarkably well-adapted features of an organism, Paley reasoned that just as the design of the former points to the existence and skills of its designer, so, too, must the apparent design in the later point to the existence and skills of its designer.

Paley's *Natural Theology* is notable not only for its persuasive articulation of the argument from design but also for its comprehensive catalogue of biological adaptations. It was a catalogue that fascinated Darwin during his studies at Cambridge. With the publication of *On the Origin of Species* (1859), he would take the same phenomena, the remarkable adaptivity of organisms, but give them a very different explanation. He would show how adaptive form can result from the differential reproduction of organisms marked by varied heritable adaptations. From the outset, it was quite plain that accepting the theory of evolution meant rejecting the argument from design. Organisms were adapted to their environment, true, but this did not necessitate their being either designed or made by an intelligent creator. And yet, as Ruse has argued (2003), Darwin did not reject the concept of design. To the contrary, he entrenched the concept of design in evolutionary theory when he famously compared natural selection to both a selective breeder and an architect. As he put it in correspondence to Hooker:

> The following metaphor gives a good view of my notion of relative importance of Variability & Selection–Squared stones, bricks or timber are indispensable for construction of a building; & their Nature will to a certain extent influence character of building, but selection I look at, as the architect; & in admiring a well-contrived or splendid building one speaks of the architect alone & not of the brick maker.
>
> (Darwin to Hooker, June 12, 1860, in Beatty 2014, 176)

In this expression, what is an architect but a designer of buildings? What is natural selection but a designer of organisms? And so, in this way, the design metaphor continued: no longer proof of God but proof of the magnificent power of natural selection. Evolution itself became the designer.

The problem, of course, is that designers are intentional, conscious, purposeful, and goal-directed. Natural selection is none of these things. What's more, designers strive to engineer optimal machines to meet their goals through efficient and streamlined processes. By contrast, natural selection is somewhat ad hoc. The process must work within the constraints of an existing system, constantly reappropriating and modifying it. If natural selection is a designer, it is a very different kind of designer from the archetype of an engineer. Out of these observations came the idea that we should perceive evolution as some kind of tinkerer (Jacob 1977), the kind of designer that constantly modifies an existing model. Alternatively, it has been argued that we can think of design without a designer (Ayala 2007), a process that is creative without being conscious. This directly refutes Paley's central claim in the argument from design that where there is design, there must be a designer. But it raises questions over exactly what design means in this context, for what is design without a designer? If by "design" biologists simply mean adaptive complexity, why not call a spade a spade? Why retain such loaded and evocative terms?

The obvious answer is that there is something truly wondrous about biological adaptations, something that has always and possibly will always fascinate biologists and philosophers alike. There are many different traits of an organism we might single out and study, but the really interesting ones, the ones we do single out, are adaptations. The concept of design helps us to emphasize what is special about these traits, what separates them from all the rest of an organism's traits: they appear designed to help an organism survive and reproduce.

We have now moved all the way from talk of an all-knowing and powerful designer God to talk of design without a designer, a concept that merely emphasizes the special status of adaptations. But even this kind of talk has its problems. As Godfrey-Smith (1999) has argued, it is this type of thinking that has resulted in much of the problems associated with adaptationism.

### Adaptationism

At heart, the adaptationist debate is concerned with the status of natural selection as a mechanism of biological change and explanation. On one side stand the so-called adaptationists, whose claims can be divided into the empirical, explanatory, and methodological (Godfrey-Smith 2001). The third type of adaptationism, methodological adaptationism, is probably the least controversial. It simply claims that we approach the study of organisms by looking for adaptations. This is a reflection of our scientific tools as opposed to a claim about the status of adaptations. By contrast, an empirical adaptationist might claim that natural selection is the most important causal power in evolution; it is subject to few constraints and, as such, organisms are simply bundles of adaptations. This is an empirical claim about the causal importance of natural selection. It is a claim that is easily refuted by, for example, considering the evolution of DNA, where it is quite clear that natural selection has played only a very minor role (Lynch 2007). While it is not clear that anyone advocates empirical adaptationism in its strict form, this view is closely aligned with the view Godfrey-Smith has termed "explanatory adaptationism." This is the view, expressed at the end of the previous section, that complex adaptivity and the apparent design of organisms are the most interesting and important questions facing evolutionary biologists; they are the proper questions for evolutionary biologists to concern themselves with. And if these are the big questions, natural selection is the big answer. It is this kind of view that seems to be held by Dawkins when he claims that the "main task of any theory of evolution is to explain adaptive complexity, i.e. to explain the same set of facts which Paley used as evidence of a creator" (1986, 404). A similar claim is made by Dennett when he claims that the only way to understand the evolution of organisms is by adopting the design stance, that is, to treat organisms as if they were designed (Dennett 1995). In the design stance, organisms are atomized into features whose function is worked out through processes of reverse engineering. Organisms are treated as if they were machines designed by evolution. The problem with this claim is that if it isn't underpinned by some kind of empirical claim, then the only motivation seems to be personal preference and interest. According to Godfrey-Smith (1999), apparent design has no special status beyond the status it enjoys in the psychology of biologists. If this is true, it raises all sorts of questions about the nature of evolution and its study by biologists, questions that have been taken up by the so-called anti-adaptationists.

Contrary to their name, "anti-adaptationists" do not reject adaptation as a mechanism for biological change. They do, however, advocate for a critical examination of adaptationist assumptions and methods (Griffiths 1996; Lewens 2004). The origin of "anti-adaptationist" thinking is largely attributed to Gould and Lewontin (1979), who argued for a pluralistic approach to evolutionary biology, one that treats organisms as integrated wholes, pays attention to their structural complexities, and considers nonselective factors in evolution. Since then, significant contributions have been made by a number of authors. Notably, Griffiths (1996) has argued that the design stance depends on "functional generalizations" that underdetermine the complexity of evolutionary processes and interspecies differences. Lewens (2004) has argued that reverse engineering suffers from limitations when applied to both organisms and artifacts. And, in general, increasing attention is being paid to the differences between organisms and artifacts as well as between evolution and engineering (Holm and Powell 2013).

One of the underlying drivers of these problems lies in a core difference between organisms and machines that the design metaphor neatly glosses over. Assuming that the only features that matter are adaptations pushes for adopting the design stance when studying organisms. The design stance assumes that an organism can be neatly decomposed into functional traits and that each trait is an adaptation, optimally designed for its function by natural selection. But, in this way, organisms are unlike machines; they cannot be as neatly decomposed into functional traits because the relationship between structural parts and functions is much more complex in an organism than it is in a machine.

## Mechanism and reductionism

The fact that an organism cannot be neatly decomposed into functional traits is often interpreted as requiring a wholesale rejection of reductionistic methods in biology (e.g., Nicholson 2012, 2013). But reductionism is not any one idea or method. To the contrary, many participants in contemporary debates allow that there might be different notions of "reduction" at play in different areas of biology (Hüttemann and Love 2011; Kaiser 2015). While certain types of reductive strategies might pack problematic ontological, epistemic, or methodological assumptions, others reveal how scientific investigation into complex biological phenomena can be fruitfully pursued. This reminds us of an important truth: the legitimacy of reductionist research strategies in different biological subdisciplines cannot and will not be solved by stipulation.

In particular, assuming near decomposability rather than complete decomposability in the investigation of organisms has proven to be a useful heuristic on many occasions. For instance, the idea has been used to account for the widespread robustness (i.e., stability in the face of change) of biological functioning. In addition, reductionist approaches have helped explain the recurrence of certain types of structures and modes of causal organization across very different kinds of organisms. Much theoretical and experimental work in biology seems to justify the expectation that dynamic yet stable biological systems are organized in nearly decomposable hierarchies (Alon 2007; Melo et al. 2016; Wagner 2005).

Methodologically, the assumption of near decomposability is also closely connected to the strategy of reverse engineering, which aims to establish how something works by investigating how its parts interact. This approach has been called "articulation of parts explanation" by Kauffmann (1970), "explanation by system decomposition" by Haugeland (1998), and "functional analysis" by Cummins (1975). More recently, a number of philosophers have made the case that reverse engineering and related heuristics give rise to "mechanistic explanations" (Bechtel and Richardson 1993/2010; Glennan 2002; Machamer 2004; Craver 2007). Mechanistic explanations of biological phenomena can focus either on how the constitutive parts of a whole are organized to produce the biological behavior or property of interest or on how the causal relations between the parts produce the behavior of the whole in a given context.

Reconstructing much of the experimental and modelling efforts of biologists in terms of the search for mechanisms and mechanistic explanations has also partially transformed how one understands the reductive agenda of biology (or at least of subfields like molecular genetics and biochemistry). Mechanistic approaches are proposed not only as promising alternatives to classical accounts of theory reduction (Brigandt and Love 2017) but also as ways of overcoming the false dichotomy between "ruthless" versions of reductionism and anti-reductionism. Thus, most mechanist philosophers reject both the idea that molecular biology can in principle explain all relevant facts about biology and the idea that some biological fields (developmental biology, ecology) are fully autonomous in the sense that they cannot benefit from theoretical ideas and findings from molecular biology (yet see Rosenberg 2006). Another alleged merit of mechanistic analyses is that they can do justice to the fact that reduction is very much still a methodological and even an explanatory goal in certain types of biological research. In line with this, some authors have developed systematic analyses of reductive heuristics and explanatory strategies encountered in scientific practice, which nevertheless succeed in avoiding the ontological idea that a reduction defines a whole in terms of its parts (see, e.g., Wimsatt 2007; Kaiser 2015). Moreover, these authors consistently acknowledge that these reductive strategies do not exhaust the methodological repertoire of practicing biologists and that their limitations are more stringent in some fields than in others.

Closely tracking biological practice, the new mechanistic framework has provided not only new analyses of explanatory reduction but also complementary models of how theoretical ideas and experimental findings are integrated among different biological fields (Darden 2006; Craver 2007). It has often been pointed out that to understand how biological phenomena come about, scientists appeal to entities at several levels of organization and that the mechanistic framework is well suited to account for the multilevel and multifield structure of explanation in biology. This double orientation of the framework (toward reduction and integration) is meant to reflect the fact that in practice scientists often change strategies depending on the research context and question asked about some target biological system.

However, ongoing scientific and philosophical debates demonstrate that the "fate" of the contemporary philosophical mechanistic framework depends on how it will manage to address the challenges that arise from treating biological systems as akin (although not identical) to machines with tightly interconnected parts that operate to produce regular outcomes. For instance, some critics have pointed out that one problem with versions of mechanistic reduction is that they cannot account for the fact that the effects of molecular processes depend on the context in which they occur, so that one molecular kind can correspond to many higher-level biological kinds. The converse challenge is to say how, from a mechanistic perspective, multiple molecular kinds can realize the same higher-level biological kind. Moreover, biological phenomena seem to be organized in temporal and processual hierarchies, as well as spatial and causal hierarchies, and mechanist philosophers have been pressed to say whether and how these aspects of biological organization are captured in a mechanistic explanatory framework. This broader set of challenges becomes even more pressing when examining the ontological/metaphysical underpinnings of the mechanistic program in the philosophy of biology.

## Process versus substance

Machines are things constituted of smaller things, each with its own properties that contribute to the machine's overall purpose. A toaster is a cube-shaped thing constituted of a heating element, a spring, a timing mechanism, and so on, all of which contribute to the toasting of bread. To understand the toaster, we can decompose it into its parts and identify their properties and how they interact. Cells, organisms, and ecosystems also appear to us as things; they have distinct forms, identifiable boundaries, and an internal organization of interacting parts. The mechanistic approach to biological study, made permissible by the machine analogy, tells us that we can understand these things by decomposition (or near decomposition) and that larger entities will be explicable by the smaller entities that constitute them. Ecosystems can be analyzed in terms of the interactions between the organisms that comprise them, organisms can be analyzed in terms of the organs that constitute them, organs in terms of the cells that constitute them, and cells in terms of the molecules of which they are constituted. But ecosystems, organisms, and cells are not just static things, they are dynamic things as well. Biological systems develop, adapt, metamorphose, and evolve. So dynamic are they that many have suggested it would be more appropriate to think of them as a series of processes rather than simply as things. John Dupré clearly expresses this sentiment when he states that we must "recognize that an organism, human or otherwise, is not a thing or substance at all, but a process. What we are looking for are not the constant properties that determine the persistence of a substance, but the activities that sustain an individual process" (2014, 8).

But this creates a problem with the machine analogy and the mechanistic approach it makes permissible. If I decompose my toaster one year and then undertake the same endeavor a year later, the resulting studies presumably will not reveal any substantial differences. Indeed, if substantial differences did exist, there would come a point where we would start to question whether the toaster was indeed the same toaster. This is the exact problem encapsulated by the ship of Theseus. Is a ship that has had all its parts replaced over time the same ship that was originally made? But these kinds of questions are simply not applicable to a biological system. To the contrary, biological systems must change the material parts of which they are made if they are to remain the same.

The centrality of change for biological systems can be seen most clearly by considering two of the most important processes for those systems: metabolism and evolution. Indeed, it was a consideration of evolution that would first motivate early process philosophers such as Whitehead and Pierce (Seibt 2018). Evolution demonstrated that the world was a constantly changing, non-static place, where novelty and innovation emerged over time. On this basis, they argued, a new metaphysics was required that prioritized processes and change over substances and essences. More recently, philosophers of biology have highlighted the role of metabolism as a process by which biological systems ensure their persistence through time by constantly repairing and replacing their material constituents, which are subject to wear and tear, and by maintaining themselves in a dynamic state. Complete stasis for a biological system means death. This is true because biological systems are open thermodynamic systems that must be maintained far from thermodynamic equilibrium (Moreno and Mossio 2015). Equilibrium means death, and organisms must engage in active processes over time to remain far from this state.

All of this points to a fundamental disanalogy between organisms and machines. While complete stasis is the ideal form of existence for a machine, an organism depends on change for its existence (Dupre 2012). This seems to suggest that organisms and other biological phenomena (cells, ecosystems) are not things at all, they are processes. If this is true, it provides another reason to question the ontological validity of the machine analogy; machines are not processes, they are the archetype of things. Given this, it has been suggested that we should scrap the machine analogy in favor of a process-oriented analogy. As Mayr puts it, "The avoidance of nouns that are nothing but reifications of processes greatly facilitates the analysis of the phenomena that are characteristic for biology" (1985, 74–75).

In fact, process analogies have long coexisted alongside the machine analogy. As Nicholson (2018) has recently made plain, biologists and philosophers have long recognized that rejecting the mechanistic approach requires finding a replacement for the machine analogy that underpins it. Cuvier suggested that an organism was like a vortex, a form that must be maintained by a constant flow of transient material. It was an idea that Whewell and Lillie later took up and advocated. Huxley compared the organism to a whirlpool, an idea that Dupré has also advocated more recently (2012). And many authors, over time (Haldane, Sherrington, Waddington, Oparin, and Bertalanffy), have compared the organism to a stream or river, drawing on the Hercalitean idea that one can never enter the same river twice, as the water it is comprised of constantly changes. It is a view that emphasizes ceaseless change and movement as opposed to the existence of atoms, the building blocks of things (Nicholson 2018).

## Intrinsic versus extrinsic purpose

One of the important features that organisms and machines seem to have in common is that their parts have functions that serve some end or purpose. It is simply not possible to explain how either an organism or a machine works without referencing those ends. But this in turn requires explanation; how do we explain this apparent purposiveness (or teleology)?

With regard to machines, this question has never caused too much trouble: Plato proposed a model of extrinsic purpose that has remained pretty much unaltered for over two thousand years. In the case of machines like clocks and cars, their purpose is imposed on them by their makers and users. The purpose of a clock is to keep the time and the purpose of a car is to transport people and other items from one place to another. Because the purpose or end to which machines and their parts function as means is determined by or originates in the interests and needs of an external agent, their purpose is said to be "extrinsic." Because it is grounded in the desires and beliefs of its maker or user, their purpose is also intentional. Finally, because it appeals to facts about the system's creation, appeals to extrinsic teleology provide ultimate explanations; they explain the origins of a system's purposiveness.

It is inarguable that organisms can have Platonic extrinsic purpose; guide dogs and generals are both examples of this by virtue of the function they serve in society. But Aristotle believed that this is not the only type of purposiveness that organisms possess. In addition to whatever extrinsic ends an organism might serve, Aristotle argued that organisms were intrinsically purposeful; they possessed a purpose that was immanent in the organism itself. It was a purposiveness that was not imposed by or beneficial to an external intentional agent. To the contrary, it arose from within the organism and benefited the organism and was usually not intentional at all. It is the Aristotelian conception of teleology that Galen would adopt in his anatomical studies, arguing that the parts of an organism must be explained by reference to their function in promoting the whole organism. This Galenic view would dominate medical thinking right up to the Enlightenment (Allen and Neal 2019).

The rise of mechanism during the Enlightenment signaled a shift away from overtly teleological explanations in favor of explanations that referenced mechanistic interactions of inert matter. But this shift was rejected by vitalists, who developed an opposing view steeped in Aristotelian biology. Although there were different forms of vitalism, they all shared the belief that physical and chemical processes were not able to explain the goal-directedness of living systems. The publication of *On the Origin of Species* in 1859 promised to offer a new naturalized explanation of the apparent purposiveness of organisms. From the outset, there was much debate as to whether Darwin had eliminated or merely transformed teleology in biology. Although this debate continues to this day, most agree that evolutionary explanations are in some sense teleological; the end for which an adaptation functions is an ineliminable feature of explanation by appeal to natural selection. But these explanations are not teleological in either the Aristotelian or Platonic sense. They are not Platonic, for they are not intentional. That said, like Platonic teleology, they provide ultimate explanations of the origins of a system's purposiveness rather than proximate explanations of how that system functions. And the ends they refer to are also external. As McLaughlin (2001) has recently argued, what natural selection explains is that functional traits that are good for the production of progeny and offspring are external ends, not internal ones. It follows from this claim that any account of functions that grounds the purposiveness of biological functions in an appeal to natural selection (e.g., Millikan 1984; Neander 1991) adopts an extrinsic conception of biological purposiveness. By contrast, the increasingly popular (among philosophers at least) organizational account of functions (e.g., McLaughlin 2001; Mossio and Bich 2017) grounds the purposiveness of functional traits within the circular organizational structure of organisms. It promotes an explicitly intrinsic conception of biological purpose in which the system is both means and end of its own activity.

Following the trajectory of this debate from Aristotle and Plato to the modern functions debate provides compelling evidence for the influence of the machine analogy. When the analogy is at its strongest, it brings with it an appeal to Platonic extrinsic purpose; when the analogy is weakened, a search for alternative models of purpose is pursued. It is important to note that this relationship is a two-way street. The current accounts of biological function base their explanations of purpose in theories of natural selection and organization. Both imply that the purposiveness of biological parts is qualitatively different from the purposiveness of machine parts. This very fact weakens the machine analogy, as it makes obsolete one of the key justifications for invoking the analogy in the first place.

Together, the themes presented in this section provide an overview of the mosaic that is the machine metaphor in biology. While it is not the case that every chapter in the volume takes up all these themes, they all relate to some of them and, as a collection, we think they will provide the reader with an overview of the current state of research into the machine metaphor and associated engineering approaches in biology.

## Aim and structure

The aim of the volume is to provide a critical assessment of what is arguably the most powerful metaphor in the biological sciences and the conceptual, methodological, and societal issues it raises. The impact of the machine metaphor in biology is certainly not waning. However, its interpretation, application, and implications are constantly being influenced by technological achievements, scientific practice, and societal developments. The chapters collected in this volume all provide insights into both historical and contemporary aspects of the machine metaphor in the scientific investigation of living systems.

The chapters are organized under three themes:

1   Theoretical issues
2   Methodological issues
3   Societal issues

### *Part 1: theoretical issues*

In her chapter, Jessica Riskin gives voice to thinkers who throughout modern history have criticized the clockwork model of nature in general and of living beings in particular. The chapter delves into the history of the struggle between what Riskin calls "passive mechanists" and "active mechanists." Both active mechanists such as Harvey, Willis, and, most notably, Leibniz and passive mechanists, which include Cartesians and Newtonians, subscribe to the view that we can understand life as the product of the components of living systems in a way analogous to the way in which we can understand the operations of a watch on the basis of the structure of its parts and their interaction. However, they disagree about whether scientific explanations should be allowed to attribute will or agency to natural phenomena, something that would make them stand out from the way in which we explain and understand the activities of man-made machines such as watches. In particular, Riskin provides a reconstruction of Leibniz's active mechanism. The central idea driving Leibniz's mechanist thinking was his dissatisfaction with understanding nature as a passive mechanical device in need of external agency for it to keep running. A naturalist and mechanist at heart, Leibniz rejected this sort of intervention and insisted that activity must come from within the natural realm. At the same time he also rejected the temptation to introduce nonmechanical principles into the living world.

Daniel J. Nicholson's chapter takes up the development of the machine conception of the organism at the beginning of the 20th century. The chapter discusses the adequacy of four machine analogies in molecular biology: "genetic program," "cellular circuitry," "molecular machine," and "molecular motor." Nicholson cautions against using these analogies due to the fact that they trade on similarities between macroscopic and microscopic entities. However, the "worlds" of these entities are fundamentally very different – the macroscopic realm is governed by gravity and inertia, the microscopic realm by Brownian motion. In particular, Nicholson is critical of molecular biologists' appeal to such analogies in explanations of the features of the objects they study. He argues that the machine analogy in molecular biology is misleading because the source and the target of the analogy exist at relevantly different scales. We should not try to understand the structure and function of molecular entities on the basis of the way we think about macroscopic electronic and mechanical machines. Against the proponents of the molecular machine metaphor Nicholson wields "the argument from scale." From the point of view of the microscopic world there are no mechanical movements corresponding to the movements of the parts of, say, a watch, and he suggests that the source of the neglect of scale when invoking the machine analogy in molecular biology is Schrödinger's argument in his famous book *What is Life?*

One of the most widely applied machine metaphors is that of "the genetic program." While Nicholson is critical of the genetic program metaphor, Brett Calcott in his chapter argues that the notion of a program has received too little attention in the debate. In particular, Calcott aims to widen the common understanding of a program as "a list of instructions telling the computer what to do." We might be able to establish a more fruitful and less misleading analogy by widening the notion of a program. Calcott's idea is that a genetic program should be understood as a form of interactive program that does not interact with an intelligent agent. To illustrate this he asks us to consider the interactive program running in a robot vacuum cleaner – a robovac. The robovac program steers the robot around its environment on the basis of input from its sensors about its local environment. Even if the robot has a single task to perform, there is no predetermined route it will use to vacuum the house. Its path around the house emerges as the result of its interaction with its environment. In "a roomful of robovacs" in which the robots adjust their behavior in response to the behavior of the others, there may emerge an organized group-level behavior such as vacuuming in formation. A multicellular organism can be said to be the result of higher-level formation of individual cells, each running what Calcott terms a "Collectively-Identical Interactive Program."

William Bechtel's contribution explores the limits of the machine metaphor in studying living organisms. He argues that the mainstream philosophical way of presenting the mechanistic framework with entities, activities, and principles of organization as key elements in explaining biological phenomena obscures important aspects of the use of the machine metaphor in biology. Bechtel proposes to extend this framework by introducing the conceptual distinction between "production" and "control" mechanisms that are together responsible for producing the regular behaviors that are the object of biological investigation. For this, he draws on the work of theoretical biologist Howard Pattee, who identifies three criteria for comparing machines to biological systems. First, Pattee points out that both machines and biological mechanisms are macroscale objects that cannot be fully characterized in terms of dynamical laws; in addition, their explanation requires the specification of constraints. Second, these constraints are set up so as to allow systems (be they machines or organisms) to perform work by limiting the flow of free energy. Finally, Pattee argues that for the ensuing work to be useful for the system as a whole, control mechanisms must operate on constraints within the production mechanisms. Bechtel emphasizes this latter distinction: in the case of machines, control mechanisms are organized hierarchically, while in the case of organisms, control mechanisms are heterarchically organized. That is, in biological mechanisms, there is not a hierarchy of controllers with a highest-level controller, but a network of heterarchically organized control mechanisms that each carry out specific control processes. The operation of "production" mechanisms needed for the organism to maintain itself is coordinated through the interaction of "control" mechanisms, not through the action of a central executive. Bechtel further explicates the distinction between machines and biological mechanisms by arguing that, in the latter case, control mechanisms operate on "flexible" constraints in production mechanisms and are sensitive to information transfers.

## *Part 2: methodological issues*

In the chapter by Arnon Levy and William Bechtel, we get a discussion of the impact a concept they call "the machine image" has had on the methods of discovering and describing biological mechanisms. In particular, they argue that the machine metaphor has been very influential in the program of "the new mechanists." Even if there has been no definitional identification of mechanisms with machine-like systems, the heavy reliance on cases that do fit the machine metaphor has arguably resulted in neglecting examination of mechanism discovery in non-machine-like systems. Levy and Bechtel then claim that New Mechanism has moved beyond the classical understanding of biological mechanisms as machine-like structures featuring clearly differentiated and spatially organized enduring components with well-defined functions working in concert to produce a specific phenomenon when start-up conditions are realized. In other words, the discovery of mechanisms in biology has in fact moved beyond machine-like systems whose behavior "can be explained by appeal to its underlying composition." Thus, Levy and Bechtel make a case for including non-machine-like systems in the class of systems that lend themselves to mechanistic investigation and explanation. A central justification for expanding the notion of mechanism to cover non-machine-like systems is that, in practice, scientists apply mechanistic analyses to such systems. Thus, Levy and Bechtel document how mechanisms are identified in systems whose behavior depends on "concentrations and related 'bulk' properties" and in systems where the parts or their organization, or both, is dynamically changing. This has important consequences for the tools and strategies adopted in research into discovering non-machine-like mechanisms.

A more general perspective on the role of analogies in science is offered in Tarja Knuuttila and Andrea Loettgers's chapter. They argue that analogy- and template-based analyses of scientific modelling can be fruitfully used to complement and augment each other. Within philosophy, analogy-based analyses focus on identifying how formal and mathematical representations are often derived by abstraction from particular target and source domains, whereas template-based analyses emphasize the generalized methods for modelling various kinds of systems. By considering the links between analogies and model templates, Knuuttila and Loettgers propose to enhance the argument for the universality of certain tractable templates. This has implications for both scientific practice and current philosophical analyses thereof. They argue that central for successful template transfer is the general conceptual core of the model template. Although model templates themselves might appear to have a purely syntactic character, Knuuttila and Loettgers argue that they also possess a conceptual dimension that is animated by analogies between various kinds of systems that are used to mobilize template transfers across a wide spectrum of domains and disciplines. It is this conceptual core, they claim, that motivates local and domain-specific template construction and adjustment processes in actual scientific practice. To illustrate their argument, Knuuttila and Loettgers discuss the application of the spin glass model in neuroscience. The Ising model of ferromagnetism originally provided the basic template for the Sherrington–Kirkpatrick model of spin glasses that in turn contributed to the Hopfield model of associative memory. This case study grounds the more general claim that both formal and conceptual elements support the inferences scientists typically make when transferring models from one domain (e.g., physics, engineering) to another (e.g., biology).

Maria Serban and Sara Green also focus on how multiple models drawn from different disciplines are used to investigate and explain biological phenomena. They argue that methods and concepts from the engineering sciences have successfully made their way into how some biologists conceptualize and classify features and behaviors of biological systems. In particular, they look into how robustness or the functional stability observed in the biological realm is understood by using control theoretic tools and models. While the success of these engineering applications to biology depends both on conceptual and empirical considerations that ground formal and informal analogies and model transfers, they nevertheless do not commit scientists to strong ontological commitments, e.g., that organisms are just like machines in the way they achieve functionally stable behaviors.

Just like Bechtel's argumentation, Serban and Green aim to extend the resources of the philosophical mechanistic framework for analyzing the modelling and inference strategies that biologists use to investigate biological phenomena. They articulate the notion of "structural-causal explanations" to account for recent engineering-based modelling efforts in systems biology that suggest that some forms of biological robustness depend on relational features of the organization of living systems. These relational features have been identified with abstract representations of patterns of organization or design principles within the framework of control theory. Serban and Green make the case that these design principles have explanatory value in the investigation of biological robustness, but they do not entail additional commitments concerning the ontological status of biological mechanisms. Instead, they maintain that the scope of the analogies on which these engineering models are based is purely methodological. To ground their argument in a concrete case study, Serban and Green study the modelling strategies developed to understand protein network robustness in the case of bacterial chemotaxis. Their analysis compares a model that interprets robustness as a result of a fine-tuning process with two related models showing that the perfect adaptation to constant stimuli by the *E coli* chemotaxis network depends on certain structural features of this network.

## Part 3: societal issues

The machine metaphor does not merely have an impact on scientific theory and methodology. It also shapes public perceptions of scientific knowledge and its possible applications. The final part of the book contains two chapters discussing the role of machine metaphors in bioethics and science communication. In his contribution, Andreas T. Christiansen examines what he calls "antimachine" views that are characterized by finding appeals to the machine-likeness of organisms to be ethically problematic. Christiansen focuses on the debate surrounding synthetic biology, a relatively young discipline that aims to introduce design principles from engineering into the biotechnological domain. Christiansen draws a distinction between the scientific and the technological machine analogy in synthetic biology. The first is characterized by highlighting the similarities in structure and workings of machines and organisms, whereas the latter aims to conceptualize the way in which biological systems can become properly engineered in terms of engineering notions such as rational design, standardization, and modularity. Christiansen finds that both of these analogies are based on a mechanistic or reductionist understanding of organisms according to which "organisms' properties and behaviors can be explained in terms of the interaction of their parts or modules (which are similarly explainable in terms of the interactions of their subparts)." In the article, Christiansen argues against both the claim that the mechanistic-reductionistic understanding causes insufficient attention to the uncertainties associated with living systems and the claim that it erodes the moral status of living beings. With regard to the first case, antimachine views overstretch the consequences of using the machine analogy. With regard to the second case, the antimachine views expand the machine analogy itself. The take-home message of Christiansen's examination of the criticism of the machine analogy in bioethics is that there is a tendency for bioethicists to broaden the scope of synthetic biologists' use of both the scientific and the technological version of the analogy.

In the final chapter, Sune Holm also focuses on synthetic biology in his examination of the role of the machine metaphor in science and science communication. Adopting Black's interactionist account of metaphor, the article discusses the heuristic and rhetorical uses of the machine metaphor. In its heuristic function, the machine metaphor inspires scientists to form new hypotheses and empirical research strategies with respect to a target phenomenon by thinking about it in terms of a phenomenon from another domain. In its rhetorical function, the machine metaphor is used to communicate complex scientific knowledge to a lay audience. In his chapter, Holm shows that synthetic biologists rely heavily on the machine metaphor as a heuristic and communicative tool. On the basis of the interactionist account of metaphor, Holm then draws attention to recent empirical studies regarding the impact metaphorical framing has on the way in which people think about an issue, and he argues that while machine metaphors might be valuable when playing a heuristic role in science, there is a real risk that their use in science communication can mislead in significant and harmful ways, e.g., by suggesting that organisms and machines are really the same kind of thing, when there are theoretically as well as practically relevant differences between them.

## Note

1  Parts of this introduction also appear in the introduction to Holt (2019).

## References

Allen, C., & Neal, J. (2019). Teleological notions in biology. In E. N. Zalta (ed.), *The Stanford Encyclopedia of Philosophy* (Spring 2020 Edition). https://plato.stanford.edu/archives/spr2020/entries/teleology-biology.

Alon, U. (2007). *An Introduction to Systems Biology*. Boca Raton, FL: CRC Press.

Ayala, F. (2007). Darwin's greatest discovery: Design without a designer. *PNAS*, 104(Suppl 1), 8567–8573.

Beatty, J. (2014). Darwin's cyclopean architect. In P. Thompson & D. Walsh (eds.), *Evolutionary Biology*, 175–192. Cambridge: Cambridge University Press.

Bechtel, W., & Richardson, R. C. (1993/2010). *Discovering Complexity Decomposition and Localization as Strategies in Scientific Research*. Cambridge Massachussetts: Princeton University Press.

Brigandt, I., & Love, A. (2017). Reductionism in biology. In E. N. Zalta (ed.), *The Stanford Encyclopedia of Philosophy* (Spring 2017 Edition). https://plato.stanford.edu/archives/spr2017/entries/reduction-biology.

Brito, C., & Marques, V. (2014). The rise and fall of the machine metaphor: Organizational similarities and differences between machines and living beings. the notion of organism. *Historical and Conceptual Approaches*, XLIII(1–4), 77–111.

Cobb, M. (2013). 1953: When genes became information. *Cell*, 153, 503–506.

Craver, C. F. (2007). *Explaining the Brain: Mechanisms and the Mosaic Unity of Neuroscience*. Oxford: Oxford University Press.

Cummins, R. (1975, November). Functional analysis. *Journal of Philosophy*, 72, 741–764.

Darden, L. (2006). *Reasoning in Biological Discoveries: Essays on Mechanisms, Interfield Relations, and Anomaly Resolution*. Cambridge: Cambridge University Press.

Darwin, C. (1859). *On the Origin of Species by Means of Natural Selection or the Preservation of Favoured Races in the Struggle for Life*. London: John Murray.

Dawkins, R. (1986). *The Blind Watchmaker*. London: Longman Scientific and Technical.

Dennett, D. (1995). *Darwin's Dangerous Idea: Evolution and the Meaning of Life*. New York: Simon ans Schuster.

Dickinson, H. (1939). *A Short History of the Steam Engine*. Cambridge: Cambridge University Press.

Dupre, J. (2012). *Processes of Life*. Oxford: Oxford University Press.

Dupre, J. (2014). A process ontology for biology. *The Philosopher's Magazine*, 67, 81–88.

Glennan, S. (2002). Rethinking mechanistic explanation. *Proceedings of the Philosophy of Science Association*, 3, 342–353.

Godfrey-Smith, P. (1999). Adaptationism and the power of selection. *Biology and Philosophy*, 14(2), 181–194.

Godfrey-Smith, P. (2001). Three kinds of Adaptationism. In S. Orzack & E. Sober (eds.), *Adaptationism and Optimality*, 335–357. Cambridge University Press.

Gould, S., & Lewontin, R. (1979). The spandrels of San Marco and the Panglossian paradigm: A critique of the adaptationist programme. *Proceedings of the Royal Society of London, Series B*, 205(1161), 581–598.

Griffiths, P. (1996). The historical turn in the study of adaptation. *The British Journal for the Philosophy of Science*, 47(4), 511–532.

Griffiths, P. (2001). Genetic information: A metaphor in search of a theory. *Philosophy of Science*, 68(3), 394–412.

Haugeland, J. (1998). *Having Thought: Essays in the Metaphysics of Mind*. Cambridge Massachusetts: Harvard University Press.

Holm, S., & Powell, R. (2013). Organism, machine, artifact: The conceptual and normative challenges of synthetic biology. *Studies in History and Philosophy of Biology and Biomedical Sciences*, 44(4), 627–631.

Holt, L. (2019). *Functional Relations Between Biological Systems*. PhD Thesis. University of Copenhagen.

Hüttemann, A., & Love, A. C. (2011). Aspects of reductive explanation in biological science: Intrinsicality, fundamentality, and temporality. *British Journal for the Philosophy of Science*, 62(3), 519–549.

Jacob, F. (1977). Evolution and tinkering. *Science*, 196(4295), 1161–1166.

Jouanna, J. (2012). The legacy of the Hippocratic treatise the nature of man: The theory of the four humours. In P. Van der Eijk (ed.), *Greek Medicine from Hippocrates to Galen: Selected Papers*, 335–360. Leiden: Brill.

Kaiser, M. I. (2015). *Reductive Explanation in the Biological Sciences*. Cham: Springer.

Kauffman, S. (1970). Articulation of parts explanation in biology and the rational search for them. *PSA: Proceedings of the Biennial Meeting of the Philosophy of Science Association 1970*, 257–272.

Lewens, T. (2004). *Organisms and Artifacts: Design in Nature and Elsewhere*. Cambridge Massachusetts: MIT Press.

Lynch, M. (2007). The frailty of adaptive hypotheses for the origins of organismal complexity. *PNAS*, 104(Suppl 1), 8597–8604.

Machamer, P. (2004). Activities and causation: The metaphysics and epistemology of mechanisms. *International Studies in the Philosophy of Science*, 18(1), 27–39.

Maturana, H., & Varela, F. (1973/1980). *Autopoiesis: The Organization of the Living*. Reidel Publishing.

Mayr, E. (1985). *The Growth of Biological Thought*. Cambridge Massachusetts: Harvard University Press.

McLaughlin, P. (2001). *What Functions Explain*. Cambridge: Cambridge University Press.

McReynolds, P. (1980). The clock metaphor in the history of psychology. In T. Nickles (ed.), *Scientific Discovery: Case Studies, Boston Studies in the Philosophy of Science*, Vol. 60. Dordrecht: Springer.

Melo, D., Porto, A., Cheverud, J. M., & Marroig, G. (2016). Modularity: Genes, development and evolution. *Annual Review of Ecology. Evolution Systematics*, 47, 463–486.

Millikan, R. (1984). *Language, Thought and Other Biological Categories*. Cambridge Massachusetts: MIT Press.

Moreno, A., & Mossio, M. (2015). *Biological Autonomy*. Dordrecht: Springer.

Mossio, M., & Bich, L. (2017). What makes biological organization teleological? *Synthese*, 194(4), 1089–1114.

Neander, K. (1991). Functions as selected effects: The conceptual analyst's defense. *Philosophy of Science*, 58(2), 168–184.

Nicholson, D. J. (2012). The concept of mechanism in biology. *Studies in History and Philosophy of Science Part C*, 43(1), 152–163.

Nicholson, D. J. (2013). Organisms ≠ Machines. *Studies in History and Philosophy of Biological and Biomedical Sciences*, 44(4), 669–678.

Nicholson, D. J. (2018). Reconceptualizing the organism: From complex machine to flowing stream. In D. J. Nicholson & J. Dupré (eds.), *Everything Flows: Towards a Processual Philosophy of Biology*, 139–166. Oxford University Press.

Oyama, S., Griffiths, P., & Gray, R. (2001). Introduction: What is developmental systems theory. In S. Oyama, P. Griffiths, & R. Gray (eds.), *Cycles of Contingency: Developmental Systems and Evolution*, 1–11. Cambridge Massachusetts: MIT Press.

Paley, W. (1836). *Natural Theology*. London: R. Faulder.

Rosenberg, A. (2006). *Darwinian Reductionism, or, How to Stop Worrying and Love Molecular Biology*. Chicago: University of Chicago Press.

Ruse, M. (2003). *Darwin and Design, Does Evolution Have a Purpose?* Cambridge Massachusetts: Harvard University Press.

Seibt, J. (2018). Process philosophy. In E. N. Zalta (ed.), *The Stanford Encyclopedia of Philosophy* (Winter 2018 Edition). https://plato.stanford.edu/archives/win2018/entries/process-philosophy.

von Bertalanffy, L. (1968). *General Systems Theory: Foundations, Development, Application*. New York: George Braziller Inc.

Wagner, A. (2005). Distributed robustness versus redundancy as causes of mutational robustness. *BioEssays*, 27(2), 176–188.

Watson, J. D., & Crick, H. C. (1953). Molecular structure of nucleic acids: A structure for deoxyribose nucleic acid. *Nature*, 171, 737–738.

Wiener, N. (1948). *Cybernetics: Or Control and Communication in the Animal and the Machine*. Cambridge Massachusetts: The Technology Press.

Wimsatt, W. C. (2007). *Re-Engineering Philosophy for Limited Beings: Piecewise Approximations to Reality*. Cambridge Massachusetts: Harvard University Press.

# Part 1
# Theoretical issues

# 1  Restless machines

*Jessica Riskin*

On a Sunday evening in November 1868, the English naturalist Thomas Henry Huxley, professor of natural history at the Royal School of Mines and of anatomy and physiology at the Royal College of Surgeons in London, friend and defender of Charles Darwin, made a joke about which people continue to chuckle almost a century and a half later, and whose humor captures what this chapter is about.

Huxley had been invited to Edinburgh by a renegade clergyman, the Reverend James Cranbook, to inaugurate a new series of "lectures on non-theological topics." Huxley chose as his nontheological topic "protoplasm" or, as he defined it for the uninitiated, "the physical basis of life." His main point was simple: we ought, he said, to be able to understand the properties of protoplasm, including its quite extraordinary property of being alive, simply in terms of its component parts, without invoking any special *something*, any force or power called "vitality" (Huxley 1869, 129).[1]

After all, Huxley pointed out – and here's the joke – water has extraordinary properties, too, but we know that it is made of hydrogen and oxygen combined in certain proportions within a range of temperatures, and we do not "assume that something called 'aquosity' entered into and took possession of the oxide of hydrogen . . . then guided the aqueous particles to their places." To be sure, Huxley continued, we do not presently understand just how water's properties follow from its composition any more than we understand how protoplasm can be alive, yet "we live in the hope and in the faith that . . . we shall by-and-by be able to see our way as clearly from the constituents of water to the properties of water, as we are now able to deduce the operations of a watch from the form of its parts and the manner in which they are put together" (ibid., 139–140).

Huxley's lecture was a huge hit. When it appeared in print as the lead article in the *Fortnightly Review* the following February, several editions of the issue sold out immediately, and John Morley, the review's editor, reckoned no article for a generation had "excited so profound a sensation" (Morley 1917, 90). The quip about aquosity continues almost a century and a half later to reappear regularly in biology textbooks and works of popular science.[2] A successful joke condenses layers of implicit argument and assumption into a very few words. In violation of the principle that one should never explain a joke (and in confirmation of the general feeling that the simpler the joke, the longer the explanation), this chapter offers an extended explanation of Huxley's quip. In particular, it addresses three aspects.

First, the joke assumes a founding principle of modern science; namely, that a scientific explanation must not attribute will or agency to natural phenomena: no active powers such as "aquosity" that "take possession" of things and "guide" them along their way. This rule also disallows, for example, explaining the falling weight driving a clock by saying that the weight wants to move closer to the center of the earth, or explaining the expansion of steam in a steam engine by saying that the steam intends to move upward toward the sky.

Second, Huxley's joke plays upon the uncertainties and hesitations involved in extending this principle of banning agency to the explanation of living phenomena, in affirming that "vitality" is no more useful or scientific a concept than "aquosity."

Finally, in place of explanations invoking mysterious powers such as "aquosity," Huxley recommended mechanist scientific explanations that took as their model of nature the workings of an artificial machine such as a watch.

This chapter briefly examines the origins and history of the principle Huxley popularized, banning agency from science, and this principle's accompanying clockwork model of nature, in particular as these apply to the science of living things. In so doing, we will also encounter a tradition of dissenters who would have rejected Huxley's punchline since they embraced the opposite principle that agency is an essential and ineradicable part of nature.

To trace the origins of modern scientific mechanism, we begin in a time and place that will perhaps be unexpected: the churches and cathedrals of late medieval and Renaissance Europe. These were thrumming with lifelike machines: automaton angels that sang and prayed, horrible devils that rolled their eyes and flailed their wings and tails, saints making ecclesiastical gestures, Christs grimacing on the cross, Virgins ascending Heavenward, and even the Holy Father himself making holy motions.

Outside of churches and cathedrals, the early modern European landscape was similarly bustling with androids and automaton animals. Sixteenth- and seventeenth-century Europeans with the means to do so established theme parks of automatic, mostly hydraulic, amusements on the grounds of their palaces and estates.

These machines engaged playfully with visitors, spraying them with water or flour or ash, hiding from or pursuing them, making faces and singing songs. Mechanism in the context of these machines did not signify the qualities it later assumed: rote, regular, constrained behavior. Instead, it signified something more like their opposites: unexpectedness, surprising behavior, responsiveness. Therefore, to propose, as the French philosopher René Descartes momentously did during the 1630s, that animal and human bodies were essentially machines did not mean that living things were passive or rote, but quite the contrary. Descartes described hydraulic grottoes in the French royal gardens in Saint-Germain-en-Laye, where automata enacted scenes from Greek mythology, in which these machines responded to spectators, fleeing from them, menacing them, engaging playfully and variously with their human visitors.

During the course of the 17th century, however, the idea of animal and human machinery narrowed into something passive, constrained, rote, without agency, really antithetical to life, mind, and spirit. For convenience, let us call the new mechanism of the later 17th century "brute mechanism." Brute mechanism actually developed, in large part, in the service of a new theological program of the 17th century – namely, arguments from design – finding evidence of a rational Designer in the rational design of his artifact, nature. Arguments from design evacuated all perception and agency to a location decisively outside the material world, leaving a fundamentally passive machinery behind. The modern scientific paradigm of nature as a complex, rational, and passive clockwork arrangement thus relied in its first instance upon a theological principle – a supernatural, rational designer-god – and mechanist scientific ideas continue to bear the imprint of this theology. For example, the theological principle of design informed the first notion of physiological adaptation during the 17th century, the idea that living beings are ideally suited to their environments, and this original theology lives on in deep disguise in current evolutionary biology.

Meanwhile, although brute mechanist natural theologians were evacuating perception and agency from nature, there were still those who struggled to hold matter, feeling, and will together to keep the machinery alive. These holdouts accordingly had something very different in mind when they talked about the "clockwork cosmos" or the "animal-machine"; let us call it "active mechanism." Active mechanism – a mechanism in which spirit and agency constituted parts of the very works – is often hard to discern through the distorting acoustics of the latter 18th and 19th centuries.

Consider how William Harvey, author of the hydraulic pump model of the heart, invoked automata in his account of the process of animal generation.

Scrutinizing the development of a chick embryo, Harvey observed that a great many things happened in a certain order "in the same way as we see one wheel moving another in automata, and other pieces of mechanism." But, Harvey said, the parts of the mechanism were not moving, as some natural philosophers claimed, in the sense of changing their places. Rather, the parts were remaining in place but transforming "in hardness, softness, colour, &ce" (Harvey 1847b, 417).[3] It was a mechanism made of changing parts.

This was an idea to which Harvey regularly returned. Animals, he thought, were like automata whose parts were perpetually transforming: expanding and contracting in response to heat and cold, imagination and sensation and ideas (Harvey 1959). The image of a mechanism of changing parts resonates with another in Harvey's treatise on the motion of the heart in which Harvey likened the heart to a "piece of machinery in which one wheel gives motion to another, yet all the wheels seem to move simultaneously" (Harvey 1847a, 31). Geared mechanisms represented constellations of motions that seemed at once sequential and simultaneous, a congress of mutual causes and effects.[4]

The first appearance of life itself, as Harvey described it, seemed to happen both all at once and as a sequence of events. Harvey wrote about seeing the chick first as a "little cloud," and then,

> In the midst of the cloudlet in question there was a bloody point so small it disappeared during the contraction and escaped the sight, but in relaxation it reappeared again, red and like the point of a pin; so that betwixt the visible and the invisible, betwixt being and not being, as it were, it gave by its pulses a kind of representation of the commencement of life.
>
> (ibid.)

A gathering cloud and, in its midst, a barely perceptible movement between being and not being. Harvey again cited both clockwork and firearms as models to depict a defining feature of this cloudy pulse that was life: the fusion of causation and simultaneity.

Elsewhere, Harvey invoked an analogy that would become commonplace by the end of the century: the relationship between an animal body and a church organ.

He suggested that the muscles worked like "play on the organ, virginals." But what Harvey meant by this comparison is striking. Later, people tended to mean a complex system of interacting parts when they compared living bodies to organs. Harvey, though, meant that the muscles performed their actions by "harmony and rhythm," a kind of "silent music" (Harvey 1959, 145–147).

These examples of ways in which Harvey invoked artificial mechanisms indicate a problem with classifying him either as a "mechanist" or otherwise; namely, that the meaning of "mechanism" was in flux. Harvey told his students at the College of Physicians that anatomy was a "mechanical" subject (Harvey 1961, 22). But what did he mean?

Well, one thing he meant was that there was no need to invoke ethereal or celestial substances in explaining physiological phenomena because the mundane elements, he said, transcended their own limits when they acted. The "air and water, the winds and the ocean" could "waft navies to either India and round this globe." The terrestrial elements could "grind, bake, dig, pump, saw timber, sustain fire, support some things, overwhelm others." Fire could cook, heat, soften, harden, melt, sublime, transform, set in motion, and produce iron itself. The compass pointing north, the clock indicating the hours, all were accomplished simply by means of the ordinary elements, each of which, Harvey said, "exceeded its own proper powers in action" (Harvey 1847b, 508–509). So his idea is not reductive but really the reverse – elevative – matter rising to new powers in action.

Similarly, Harvey elsewhere defined "mechanics" as "that which overcomes things by which Nature is overcome." His examples were things having "little power of movement" in themselves that were nonetheless able to move great weights, such as a pulley. Mechanics, understood in this way, included natural phenomena that overcame the usual course of nature: Harvey again mentioned the muscles. So to say that the muscles worked *mechanically* in this instance meant that the muscles, like artificial devices such as a pulley, overcame the usual course of nature and moved great weights without themselves being weighty (Harvey 1959, 127). Again, here is a form of mechanism that is really the opposite of reductive.

Another way to get at what "mechanism" meant to Harvey is to look at what posed problems for a mechanist physiology and what sorts of solutions Harvey proposed. One problem he identified, for example, was action at a distance. He was working from the Aristotelian view that embryos arose from a kind of contagion, "a vital virus" with which the sperm infected the egg.[5] And Harvey arrived at the problem of action at a distance. After the initial moment of contact, once the contaminating element had disappeared and become "a nonentity," he wondered, how did the process continue? "[H]ow, I ask, does a nonentity act? How does a thing which is not in contact fashion another thing like itself?" (Harvey 1847b, 359–360). Aristotle, Harvey pointed out, had invoked automata, automatic puppets, to explain this. He had suggested that the initial contact at conception set off a succession of linked motions that constituted the development of the embryo (Aristotle, *Generation of Animals*, Bk 2). Harvey rejected this explanation (ibid., 345–346).[6]

In its place, he proposed a different analogy: one between the uterus and the brain.

He observed that the two were strikingly similar in structure and reasoned that "where the same structure exists there must be the same function." The functions of each were called "conceptions," and perhaps they were essentially the same sort of process (Harvey 1847b, 372, 577–579, 585). A brain produced works of art by bringing an immaterial idea to matter. Perhaps a uterus produced an embryo in the same way, by means of a "plastic art" capable of bringing a form to flesh. The form of an embryo would then exist in the uterus of the mother just as the form of a house existed in the brain of the builder. And this would solve the problem of action at a distance. The moment of insemination endowed the uterus with an ability to conceive embryos in the same way that education endowed the brain with the ability to conceive ideas. Once the seed disappeared, it no longer needed to act: it was the uterus itself that took over the task of fashioning the embryo (ibid.).

So, the idea that the uterus functioned like a brain, actively fashioning an embryo the way a brain fleshes out an idea, was for Harvey not only within the bounds of the "mechanical" but also a model that could in fact rescue mechanism by eliminating the need for action at a distance.

Another person who, like Harvey, resisted the central tenets of brute mechanism was Thomas Willis, early cartographer of the brain and nervous system. Willis, like many people, disliked Descartes's description of animals as automata, but he did not reject the idea of animal machinery. Rather, Willis took Descartes to have meant that animals were purely passive, moving only when set in motion by impacts on their material souls from outside (Willis 1683, 3).[7] (Descartes himself, as we have seen, thought of machines as active and responsive things. But by the end of the 17th century, most people nevertheless construed his idea that animals were automata, as Willis did, to mean that animals were purely passive.) Willis

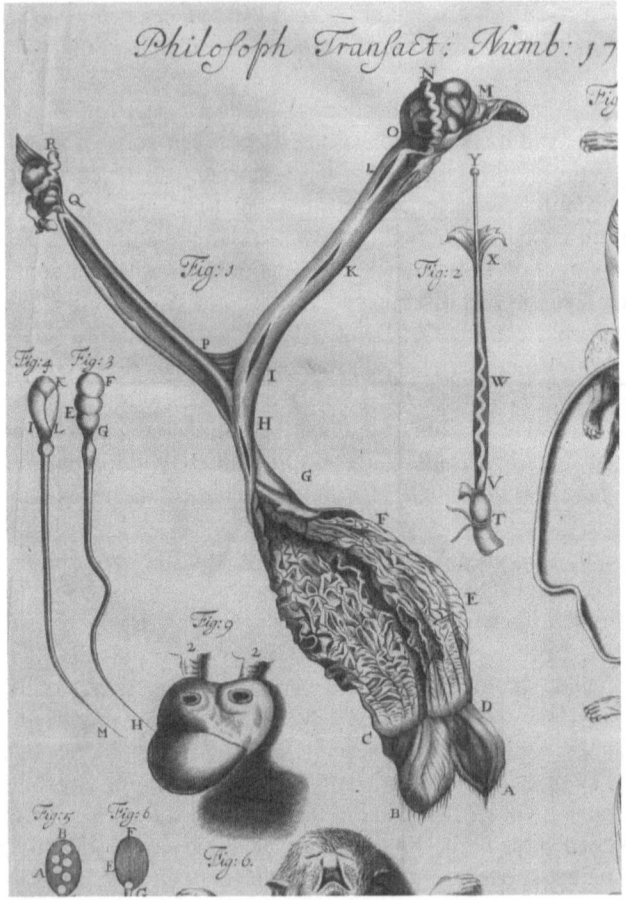

*Figure 1.1a and 1.1b* Leeuwenhoek's dog uterus and Willis's brain.

Source: Courtesy of the Department of Special Collections, Stanford University Libraries.

Note: This figure can be accessed in color via the eBook version of the book and eResources at www.
routledge.com/9780815380788.

disliked the passivity of this view of the animal-machine. He described instead a
"self-moving," active animal-machine (ibid., 56).

Like Descartes's account of animals, Willis's was fully materialist.

The animal soul he described was coextensive with the body and made of particles of the same matter, but the choicest among these, the most "subtle and highly
active." These "nimble" particles directed the formation of the animal body, gathering together in "Turgid" heaps, jostling, stirring, and steering one another and
the grosser particles to their proper places (Willis 1683, 6). Willis's animal soul
had physical parts and members, "Pipes and other Machines," and this rigorously

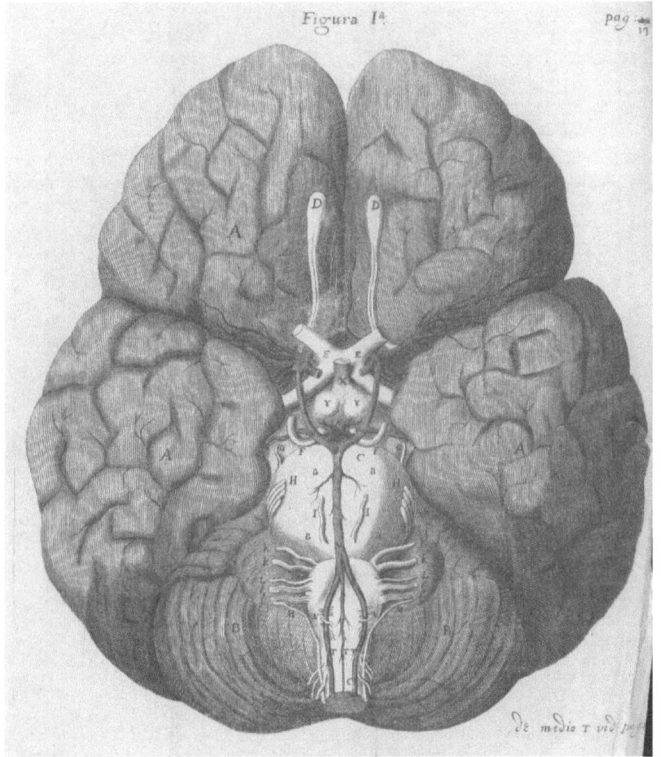

*Figure 1.1* (Continued)

material entity was "Knowing and Active," even capable of learning (ibid., 24 and 34).

Artificial machinery offered Willis plenty of models on which to base his idea of a vital, perceptive, active animal-machine. He went beyond the clock-maker's assemblage of wheels and gears, which was, after all, only a narrow domain in the growing expanse of human-made devices. Willis pointed out that "mechanical things" required "Energetical" components: "Fire, Air, and Light." Any smith, chemist, glassmaker, lens grinder, or instrument maker could eas-ily testify to the truth of this. Animal souls, in the same way, were made of the most energetic particles of matter. The movement of these particles through the animal body was like "a blast of Wind" in a wind-driven "Machine," running "hither and thither" and producing all the animal's sensations and movements (ibid., 24 and 56). Willis's drawing of the nerves of the trunk does strikingly resemble an organ.

But when Willis, like Harvey, drew the standard analogy between animals and hydraulic organs, he again meant something distinct by it. Looking at an

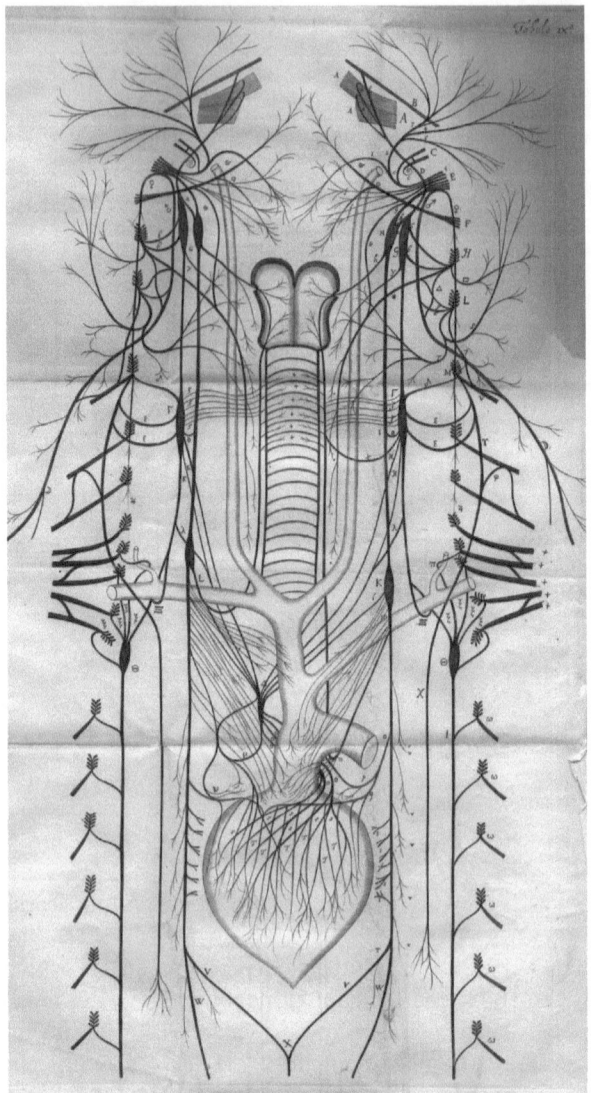

*Figure 1.2* Willis's drawing of the nerves.

Source: Courtesy of the Department of Special Collections, Stanford University Libraries.

Note: This figure can be accessed in color via the eBook version of the book and eResources at www.routledge.com/9780815380788.

automatic organ, he saw constraint, but not passivity. The "Soul of a Brute," he said, was like a "water Organ": it could play only a limited repertoire of tunes, but nevertheless, in doing so, it could actively "Institute, for ends necessary for its self, many series of Actions" (ibid., 34).

The person who probably traveled the farthest toward establishing mechanist science along a different trajectory was the German rationalist philosopher G. W. Leibniz.

He insisted that everything in nature happened mechanically (Leibniz 2004, 105), while at the same time he rejected brute mechanism, finding it to be a poor description not only of animals but also of machines (Leibniz 1994b, 197).

Leibniz was among the many who denied Descartes's claim that animals lacked immaterial souls. But in his case, it was part of a more general refutation of Cartesian physics: to Leibniz, nothing really lacked a soul. No sort of brute machinery, no matter how subtle, could account for perception, but neither could it account for any sort of action or motion (Leibniz 1710). Leibniz found Descartes's version of clockwork to be implausible even as an account of a clock, let alone of a dog, so that while others argued that animals were more than machines, Leibniz insisted that machines were more than machines. Matter drained of spirit could explain nothing. Mechanics required something more, Leibniz said – namely, force, energy, modes of action (Leibniz 1994a, 66).

This view of Leibniz's is best known in the context of the so-called *vis viva* controversy of the 1680s and '90s in which Leibniz defended his notion of a conservation of *vis viva* or "living force" (mathematically equivalent to our kinetic energy) against Cartesians who argued for Descartes's principle of conservation of motion (equivalent to our momentum).[8] Motion, Leibniz argued, was "not something entirely real" because it consisted merely of relations among objects, whereas force, a "force of acting," was "something real," belonging to a given body in itself (Leibniz 1989a, 51).[9] Moreover, no one, he said, had ever explained force (ibid., 119 and 123), and he rejected what he saw as the current tendency to instead "summon God *ex machina*, and withdraw all force for acting from things" (ibid., 125).

On the other hand, Leibniz also disliked the idea of an "Archaeus," or Paracelsian vital spirit, which he found "unintelligible": "as if not everything in nature can be explained mechanically, and as if those who try to explain everything mechanically are thought to eliminate incorporeal things" (ibid., 126). Rather, Leibniz was after a third way, neither marionette-mechanism with God as the puppeteer – such that "when a person thinks and tries to move his arm, God moves the arm for him," an idea so absurd it "ought to have warned these writers they were depending on a false principle" (ibid., 130) – nor alchemical abandonments of mechanism, but a fully mechanist account of nature that incorporated immaterial, "active force" (Leibniz 1860).

Leibniz's third way was not purely speculative but became a model for an alternative approach to physics and engineering. His notion of *vis viva* informed a tradition growing through the 18th century and into the 19th that culminated in the development of an energistic physics and the concepts of energy and work.

Everywhere in material events, Leibniz found an active agency working from within: "a flare that runs the length of a cord or a fluid that runs in a channel." The laws of mechanics required "active indivisible principles" to make things go (Leibniz 1994b, 194). In Leibniz's world-machine, perception and agency were thus not banished but active within the very works.

When he called a process "mechanical," Leibniz therefore did not mean that it was drained of spirit, agency, and perception. Rather, he meant that it followed entirely from its own internal principles, with no appeal to a *Deus ex machina*.[10] Whereas naturalism and mechanism to some meant strictly banishing purpose and agency from the works, what Leibniz called a "purely natural and entirely mechanical" (Leibniz 1989g, 344) account of things meant, quite the contrary, a *naturalizing* agency – understanding the world in terms of a fundamentally purposeful and active kind of machinery. In this sense, the two kinds of mechanism are really antithetical.

A thoroughly naturalist, mechanical theory of nature, according to Leibniz, would need to include the ultimate causes as well as the proximate ones; it must encompass the metaphysical principles governing force and the laws of motion. And, of course, Leibniz's great objection to the Newtonian system, which he voiced emphatically during the last couple years of his life in his epistolary debate with Samuel Clarke, was that Newton described the cosmos as an artifact, a device of brute mechanism. Like any such device, it required its Maker to step in and adjust it, to rewind it and keep it running. The very running of Newton's cosmic Clock, therefore, rested upon an extra-natural cause, an actor intervening from outside the system. In Leibniz's view, this was a violation of the principles of naturalism and mechanism.

According to his ideal of science, there should be one single system encompassing all of nature, with nothing outsourced, nothing rendered exceptional or external.

Regarding the authors of the standard mechanical philosophy – Galileo, Descartes, Hobbes, Gassendi – Leibniz remarked that they had "purged inexplicable chimera from philosophy," but at a high price: they had exported the chimera, leaving a metaphysical gap that they filled with a meddlesome God who acted supernaturally. "I," wrote Leibniz, "tried to fill this gap, and have at last shown that *everything happens mechanically in nature, but that the principles of mechanism are metaphysical*" (Leibniz 1989e, 318). That is, the mechanism of nature is intrinsically purposeful.

In reacting against the brute mechanism of his contemporaries, Leibniz included its insufficiency to account for life and mind. He appealed to internal experience, the inner consciousness of "this *me*" that he thought would never be explicable by "figures or movements" (Leibniz 1702, 197). And in a famous thought experiment designed to show this, Leibniz instructed his reader to imagine a big machine, the size of a mill, that could think, feel, and perceive. Imagine, he wrote, walking into this great factory of thought and looking around. You would find only "pieces that push each other and never anything to explain a perception." You would understand consciousness no better than before entering the mill of mind. Looking at the machinery, the pushing and pulling, the moving parts, the thing you would be led to understand was that perception and consciousness were not that. Perception, Leibniz wrote, resided not in the operation of the mechanism but in its very substance. Indeed, perception *was* the primary substance: mind was the *stuff* of the machine (more on that presently) (Leibniz 1996, 17).

So Leibniz included the inability of moving parts to explain perception and consciousness. But this was part of a more general insistence on incorporating rather than reducing or exporting these phenomena. While he rejected brute mechanism, he equally rejected the opposite extreme, the organicist view that the physiological processes of living creatures were essentially different from inanimate phenomena. All of nature, Leibniz thought, should be explicable by the same mechanical principles.

If all things in nature happened mechanically, it meant a mechanical philosophy quite different from the other reigning models, such as the Cartesian or the Newtonian model, or that of Robert Boyle. Over the course of his career, Leibniz developed an increasingly distinct understanding of mechanism[11] until, in the *Monadology* (1714), he reasoned that the most basic elements could not be material bits because any corporeal entity – no matter how "invincibly attached" – must be divisible. One could still imagine dividing it further. Therefore, the indivisible atoms that made up the world must be something else. And the something else he settled upon, to replace extension, was perception. Rather than being extended bits, which Leibniz thought ultimately explained nothing in nature, the most basic elements of the world were perceiving souls (Leibniz 1994a, 71). And by substituting perception for extension, Leibniz offered an equal and opposite philosophy to Hobbesian materialism: a reduction of matter to spirit, with perceiving spirits in the place of chunks of matter.

The cosmos and everything within it, including animals and humans, was a great nesting of machines within machines within machines, all built out of little perceiving spirits. Natural machinery, Leibniz wrote, was mechanical down to its "smallest distinguishable parts" (Leibniz 1989f, 207).[12] Here was the core principle in his development of a theory of organisms in the first decade of the 18th century.[13] He wrote in 1704:

> I define the Organism, or the natural Machine, [as] a machine in which each part is a machine, and consequently the subtlety of its artifice goes to infinity, nothing being small enough to be neglected, whereas the parts of our artificial machines are not machines. That is the difference between Nature and Art.

Living nature was machinery all the way down. Our body, therefore, Leibniz continued, "is a sort of world full of an infinity of creatures who also deserved to exist" (Leibniz 1875, 356). He later presented this idea in its full form in the *Monadology*:

> Thus each organic body of a living being is a sort of divine machine, or a natural Automaton, that infinitely surpasses all artificial automata. Because a machine made by the art of man is not machine in each of its parts, for example the teeth of a brass wheel have parts or fragments that are not artificial things in our view and have nothing to mark them as machines in relation to the usage for which the wheel was destined. But natural machines, that is, living bodies, remain machines in their least parts to infinity.

(Leibniz 1996, 64)

This cosmos of machines within machines without end, their ultimate components living souls, was intrinsically active, self-moving, containing its own energies, forces, powers of action. Leibniz presented this view that things in nature contain their own energy and forces in explicit response also to Robert Boyle's mechanical cosmos (Leibniz 1969, 498–508). Like Descartes and Newton, Boyle drained the cosmic machinery of agency. Leibniz's cosmic machinery, in contrast to Boyle's, was brimming with life and sentience in every part. And this lively, sentient mechanical nature was of course permeated to the core with perception, too, being made of perceiving spirits (Leibniz 1996, 14). Each speck was a garden of plants, each drop a pond teeming with fish. Every part of every plant or animal and all the air between them was lush with living creatures. "All of nature," Leibniz wrote, "is full of life" (Leibniz 1989f, 207) – it was in fact exactly as full of life as it was of mechanism.

Having met the Dutch microscopists Swammerdam and Leeuwenhoek in 1676 during the same visit to the Netherlands when he met Spinoza,[14] Leibniz cited microscopic evidence in support of his claim that the tiniest particle of matter contained whole worlds of living machines and that these were intrinsically active (Leibniz 1996, 74). Swammerdam, for example, described how a cuttlefish sperm, upon being removed from the cuttlefish gonad, began to act, and he ascribed full agency to what he called the "minute machine" itself.

Moreover, the self-directed movements of this minute machine were not ticks and tocks but twists and turns, evolutions, unfoldings, emergences: "[t]he extremity begins to evolute and unfold itself," Swammerdam wrote, "and the two slender ligaments, which emerge out of the case, turn and twist themselves in various directions" (Swammerdam 1758, 147).

What Leibniz meant by "machinery," then, was not just different from but in a sense opposite to what others such as the Cartesians, Hobbes, Newton, and Boyle all meant. There was no actual pushing and pulling, no action by impact, but only the appearance of these mechanical causes. Indeed, matter itself was an appearance, a secondary effect of the perceptual substance out of which the world was composed. Monads could not change one another, having no extension, no parts (as Leibniz famously described it, no "windows" through which anything could come in or go out). Rather, each little soul followed its own internally directed sequence of changes that had been set in motion at the beginning of time, and a preestablished harmony coordinated all these sequences so as to follow the laws of mechanics (Leibniz 1996, 1–15).

When Leibniz, like everyone else, compared God to a watchmaker and his creations to automata, or when he wrote that the bodies of men and animals were "no less mechanical" than watches (Leibniz 1989d, 345), he meant something profoundly different from the very same statement by, say, Descartes or Boyle.

To see how very different his meaning was, consider the following passage from the *New Essays*:

> In German, the word for the balance of a clock is *Unruhe* – which also means disquiet; and one can take that for a model of how it is in our bodies, which

can never be perfectly at their ease. For if one's body were at ease, some new effect of objects – some small change in the sense-organs, and in the viscera and bodily cavities – would at once alter the balance and compel those parts of the body to exert some tiny effort to get back into the best state possible; with the result that there is a perpetual conflict which makes up, so to speak, the disquiet of our clock; so that this [German] appellation is rather to my liking.

(Leibniz 1690–1702, 164)

Nowhere in this passage, among the metaphorical significations of "clock," do we find what had by then become, and still remain, the expected connotations: regularity, imperturbability, precision. Instead, we have something like their opposites: disquiet, unease, exertion, conflict.[15] Wherever Leibniz returned to the notion of an animal as a machine his meaning was similarly exotic. He described natural, animal machinery as "entangled," waxing and waning, enfolding and unfolding, "frail" and yet capable of self-maintenance (Leibniz 1989c, 253).

Active and passive mechanism remained embroiled in competition for the life sciences through the 18th century and into the 19th. You can see this competition in the explosion of projects to build machines that could eat, talk, play musical instruments, write, draw, and perform other human tasks, and especially in the conversation around these new experimental machines, in the Enlightenment man-machine thought experiment that commanded so much interest and provided the foundation for thinking about philosophical and social questions of every kind.

And you can see the development of passive mechanism particularly in the refinement and application of the argument from design in its application to physiology. This tradition reached a pinnacle in William Paley's natural theology, with his famous image of the watch on the heath, the text that Darwin memorized as a student at Cambridge. If you tripped over a stone while crossing the heath, you might suppose it had always just been there, Paley reasoned, but if you tripped over a watch, you could hardly make the same supposition. By the same token, the beautiful and rational living clockwork of nature points the way to a divine watchmaker. This argument of course assumes that the watch is as passive as a stone, lying on the heath, waiting to be kicked aside. A moving, self-constituting, and self-transforming sort of watch, such as the active mechanists considered living beings to be, would hardly carry the same implication.

The tradition of active mechanism bore a momentous result over the course of the 18th century. The insistence that mechanism encompass purpose and agency set the natural machinery of the world in forward motion. Because the order in nature came within and through its purposeful machinery, rather than being delivered all at once from outside at the beginning of time, this meant it had to occur in natural time: it must unfold. This sort of order, in other words, was not "design" but "organization," not a static structure but a patterned process. Passive, inert mechanism could not by itself account for the formation and development of life, Leibniz had argued, but mechanism in motion – mechanism endowed with a plan – could fully do so (Leibniz 1989a, 40–42).

Accordingly, everything in Leibniz's lively world was in a perpetual state of flux. His cosmos flowed like a river. His natural machines were always developing and changing, like the "spermatic machine" developing into the embryonic machine (Leibniz 1875), enfolding, extending, contracting. No natural machine ever appeared or disappeared, Leibniz wrote. When it seemed to be gone, it was merely "concentrated" (Leibniz 1994a, 70–71). Living machines had no true beginnings, apart from the beginning of time, and no real endings: births and deaths were mere appearances (ibid., 68). In reality, there was only development and growth, envelopment and diminution. And in this continual waxing and waning, certain souls rose "to the degree of reason and to the prerogative of minds" (Leibniz 1996, 71, 73, 76, 82).

This account of an unfolding, organic order in nature closely informed the ideas of species change that began to emerge around the middle of the 18th century and, through them, the first full-fledged theory of species change of Jean-Baptiste Lamarck around the turn of the 19th century.

Meanwhile, the brute-mechanist tradition of arguments from design provided a locus for the emergence of notions of adaptation and fitness in physiology. So here were the two principal components of early evolutionary theories, the idea of species change and the idea of adaptation and fitness, coming out of the active- and brute-mechanist traditions, respectively.

This left a kind of contradiction at the heart of evolutionary theory. It inherited from argument-from-design natural theology, packaged with the concept of adaptation, a brute mechanist approach to living structures, one that banned all reference to agency within natural phenomena. At the same time, evolutionary theory also inherited a diachronic order that had grown out of an active mechanist understanding, not only of living creatures but also of nature overall, which was really at odds with brute mechanism.

The tension between classical and active mechanism maintains a powerful subterranean activity in current biology. The evolutionary concept of adaptation encapsulates this problem: it casts creatures as clockwork, in the sense of a mechanical fitness of parts, and yet with no remaining sense that a clock could be in any sense a restless and disquiet thing, in the perpetual state of responsiveness, flux, and conflict that is life.

## Notes

1 On James Cranbrook and his radical lecture series, see Stratham, "The Real Robert Elsmere."
2 Some recent examples: Hunter, *Vital Forces*, p. 70; Wayne, *Plant Cell Biology*, p. 5; Berkowitz, *The Stardust Revolution*, p. 120.
3 On page 346, Harvey cites Aristotle, saying essentially the same.
4 In the same way, Harvey also found the heart to be like the mechanism in firearms, in which the sequence of trigger, flint, steel, spark, powder, flame, explosion, and shot all seemed to take place "in the twinkling of an eye," W. Harvey, *An Anatomical Disquisition on the Motion of the Heart* [1628] (1847), pp. 31–32.
5 The passage in Aristotle to which Harvey referred was *History of Animals*, Bk. 6, Part 13; W. Harvey, "*Anatomical Exercises on the Generation of Animals*, to which

are added essays on parturition; on the membranes, and fluids of the uterus; and on conception," in Robert Willis (trans.), *The Works of William Harvey, M.D.* (London: Sydenham Society, 1651), p. 359.

6 See also *De motu locali animalium* (ca.1627), 147 and 151, where Harvey canvasses the various models of order: a well-governed state, the work of masons, bricklayers, and carpenters; the working of a ship, an army, and a choir.

7 Willis attributed the same view to the Spanish doctor and philosopher Gomez Pereira before Descartes and to Kenelm Digby afterward.

8 On the *vis viva* controversy, see Carolyn Iltis, "Leibniz and the Vis Viva Controversy," *Isis* 62, No. 1 (Spring 1971), pp. 21–35; and Daniel Garber, "Leibniz: Physics and Philosophy," in Nicholas Jolley, ed., *The Cambridge Companion to Leibniz* (Cambridge: Cambridge University Press, 1995), Ch. 9, pp. 309–314.

9 Leibniz first presented his refutation of Descartes's principle of conservation of motion in "Brevis demonstratio erroris memorabilis Cartesii. . .," in *Acta Eruditorum*, 1686, pp. 161–163; Leibniz, *Specimen Dynamicum* (1695), in R. Ariew and D. Garber, eds., *Philosophical Essays* (Indianapolis: Hackett, 1989), pp. 117–138.

10 See, for example, Leibniz, *Système nouveau* (1695), p. 72; and "Postscript of a Letter to Basnage de Beauval (1696)," in Ariew and Garber, eds., *Philosophical Essays*, pp. 147–148.

11 On this trajectory in Leibniz's thinking, see especially Duchesneau, *Les Modèles du vivant* (1998), pp. 315–372.

12 See also Leibniz, *Système nouveau* (1695), pp. 70–71: "A natural machine still remains a machine in its least parts"; and *Nouveaux essais sur l'entendement humaine* (ca. 1690–1705), Carl Immanuel Gerhardt, ed., *Die philosophischen Schriften von Gottfried Wilhelm Leibniz* (Berlin: Weidmannsche Buchhandlung, 1875) **[hereafter PS]**, Vol. 5, pp. 40–509, Bk. 3, Ch. 6, §39.

13 See Duchesneau, *Les Modèles du vivant* (1998), Ch. 10 and "Leibniz's model for analyzing organic phenomena" (2003), p. 398. Duchesneau argues that Leibniz's theory of organisms grew out of his understanding of dynamics, a "science of power and action," which he developed in the last decade of the 17th century.

14 On Leibniz's encounter with Jan Swammerdam, see Stuart Brown, "The Seventeenth-Century Intellectual Background," in Jolley, ed., *The Cambridge Companion to Leibniz*, Ch. 3, p. 63, fn. 41. Leibniz, *Monadologie* (1714), p. 74. On his meeting with Anton von Leeuwenhoek, see R. S. Woolhouse and Richard Francks, eds. and trans., *Leibniz's 'New System' and Associated Contemporary Texts* (Oxford: Oxford University Press, 1997), p. 13, n.35. On both, see also Roger Ariew, "G.W. Leibniz, His Life and Works," in N. Jolley, ed. (Cambridge: Companion, 1995), Ch. 2, p. 27.

15 Leibniz was responding in this passage to Locke's discussion in the *Essay on Human Understanding*, Bk 2, Chs 20–21, of pleasure and pain, and building Locke's notion of "uneasiness" into a physical yet non-reductive theory of the basis of human action and behavior. The key element in this transition is Leibniz's notion of unconscious perceptions, which makes a continuity between bodily and conscious responses without reducing either to deterministic mechanism.

# References

Aristotle. *Generation of Animals*, Bk. 2, 734b3–734b18, 741a32–741b24, see also Chapter Three.

Berkowitz, Jacob. *The Stardust Revolution: The New Story of our Origin in the Stars*. Amherst, NY: Prometheus Books, 2012.

Duchesneau, François. *Les modèles du vivant de Descartes à Leibniz*. Paris: J. Vrvin, 1998.

Harvey, W. (1847a [1628]). *An Anatomical Disquisition on the Motion of the Heart*, Vol. 31.

Harvey, W. (1847b [1651]). Anatomical exercises on the generation of animals, to which are added essays on parturition; on the membranes, and fluids of the uterus; and on conception. In Robert Willis (trans.), *The Works of William Harvey, M.D.* London: Sydenham Society.

Harvey, W. (1959 [1627]). *De motu locali animalium.* Gweneth Whitteridge (ed. & trans.). Cambridge: Cambridge University Press.

Harvey, W. (1961 [1616]). *Lectures on the Whole of Anatomy* [*Prelectiones Anatomiae Universalis*]. C. D. O'Malley, F. N. L. Poynter, & K. F. Russell (eds. & trans.), 22. Berkeley: University of California Press.

Hunter, Graeme K. Vital forces: *The Discovery of the molecular Basis of life.* London: Academic Press, 2000.

Huxley, T. H. (1869). *On the Physical Basis of Life.*

Leibniz, G. W. (1690–1705 [1710]). *Reflections on the Souls of Beasts.* Donald Rutherford (trans.), *New Essays,* Bk. 2, Ch. 20, 164.

Leibniz, G. W. (1860, Halle). "Essay de dynamique sur les loix du mouvement, où il est monstré, qu'il ne se conserve pas la même quantité de mouvement, mais la même force absolue, ou bien la même quantité de l'action motrice." In C. I. Gerhardt (ed.), *Mathematische Schriften,* 9 vols, Vol. 6, 215–231.

Leibniz, G. W. (1875–1890 [30 June 1704]). Lady Damaris Masham. In C. I. Gerhardt (ed.), *Die philosophischen Schriften von Gottfriend Wilhelm Leibniz,* 352–357. Berlin: Weidmann.

Leibniz, Remond 11 february 1715, in Leibniz. *Die philosophischen Schriften* (1875–1890), 3: 635.

Leibniz, G. W. (1969 [1698]). *De ipsa natura.* in English translation as *on Nature Itself, or on the Inherent Force and Actions of Created Things.* In Leroy E. Loemker, ed. trans., *Philosophical Papers and Letters.* Dordrecht: Reidel, 1969, 498–508.

Leibniz, G. W. (1989a [1686]). "A Discourse on Metaphysics." In R. Ariew & D. Garber (eds.), *Philosophical Essays,* 35–68. Indianapolis: Hackett.

Leibniz, G. W. (1989b [1695]). "Specimen Dynamicum." In R. Ariew & D. Garber (eds.), *Philosophical Essays,* 117–138. Indianapolis: Hackett.

Leibniz, G. W. (1989c [May 1702]). "On body and force, against the Cartesians." In R. Ariew & D. Garber (eds.), *Philosophical Essays,* 250–256. Indianapolis: Hackett.

Leibniz, G. W. (1989d [1710]). *Essais de théodicée.* In R. Ariew & D. Garber (eds.), *Philosophical Essays,* 320–346. Indianapolis: Hackett.

Leibniz, G. W. (1989e [1710–16]). *Toward a philosophy of what there actually is and against the revival of the qualities of the scholastics and chimerical intelligences.* In R. Ariew & D. Garber (eds.), *Philosophical Essays,* 312–319. Indianapolis: Hackett.

Leibniz, G. W. (1989f [1714]). "Principles of nature and grace based on reason." In R. Ariew & D. Garber (eds.), *Philosophical Essays,* 206–213. Indianapolis: Hackett.

Leibniz, G. W. (1989g [1715–1716]). *Letters to Clarke.* In R. Ariew & D. Garber (eds.), *Philosophical Essays,* 320–346. Indianapolis: Hackett.

Leibniz, G. W. (1994a [1695]). *Système nouveau de la nature et de la communication des substances.* In Christiane Frémont (ed.), *Système nouveau de la nature et de la communication des substances et autres textes,* 66. Paris: Flammarion.

Leibniz, G. W. (1994b [1702]). "Réponse aux réflexions continues dans la seconde edition du Dictionnaire critique de M. Bayle, article Rorarius, sur le système de l'harmonie préétablie." In Christiane Frémont (ed.), *Leibniz, Système nouveau de la nature et communication des substances et autres textes,* Ch. 13, 194, 197–198. Paris: Flammarion.

Leibniz, G. W. (1996 [1714]). *Monadologie*. Leibniz, *Principes de la philosophie [Monad-ologie]*. In Christiane Frémont (ed.), *Principes de la nature et de la grâce, Monadologie et autres textes*, 17. Paris: Flammarion.

Leibniz, G. W. (2004 [1709]). "Repliques aux observations de Stahl." In Carvallo (ed.), *La Controverse entre Stahl et Leibniz sue la vie, l'organisme et le mixte*, 101–144, Réplique 2. Paris: J. Vrin.

Leibniz, R. (1875–1890 [11 February 1715]). *Die philosophischen Schriften*, Vol. 3, 635.

Morley, V. J. (1917). *Recollections*, Vol. 1. New York: Macmillan Company.

Stratham, F. Reginald. "The Real Robert Elsemore." *National Review* 28, 164 (October 1896): 252–61.

Swammerdam, J. ([1676–1679] 1758). *The Book of Nature, or, The History of Insects*. Thomas Flloyd (trans.). London: C.G. Seyffert, Part II, 147 fn.

Wayne, Randy O. *Plant cell Biology: from astronomy to zoology*. San Diego, CA: Elsevier, 2009.

Willis, T. (1683). *Two Discourses Concerning the Souls of Brutes*. London.

# 2  On being the right size, revisited

## The problem with engineering metaphors in molecular biology

*Daniel J. Nicholson*

In 1926, Haldane published an essay titled *On Being the Right Size* in which he argued that the structure, function, and behavior of an organism are strongly conditioned by the physical forces that exert the greatest impact at the scale at which it exists. This chapter puts Haldane's insight to work in the context of contemporary cell and molecular biology. Owing to their minuscule size, cells and molecules are subject to very different forces than macroscopic organisms. In a sense, macroscopic and microscopic entities inhabit different "worlds": the former is ruled by gravity and inertia, whereas the latter is governed by Brownian motion. One implication is that we should be extremely skeptical of models and analogies that seek to explain properties of microscopic entities by appealing to properties of macroscopic ones. Unfortunately, this is precisely what the appeal to engineering metaphors in molecular biology attempts to do. Molecular biologists routinely resort to such metaphors because they are familiar and intuitively intelligible. But if our machines were the size of molecules it would be impossible for them to function the way they do. It follows that we should avoid distorting biological reality by construing it in engineering terms. In this chapter I examine four key metaphors in molecular biology – "genetic program," "cellular circuitry," "molecular machine," and "molecular motor" – and I argue that their deficiencies derive from their neglect of scale. I also try to explain why many biologists today appear to have forgotten the importance of scale that Haldane drew attention to in his essay. I suggest that the reason has to do with the influence of Schrödinger's argument in *What is Life?* regarding the stability of the gene.

## Introduction

Machines have been used as sources of metaphorical and analogical explanations for as long as organisms have been the subject of empirical investigation. Aristotle compared the bones and tendons of the forearm to the arms of a catapult drawn back by tightening ropes. Descartes was so impressed by the life-like movements of hydraulic automata that he concluded that the movements of the body were machine-like. And Liebig's view of digestion as combustion relied on his understanding of the body as a heat engine and of food as fuel. In more recent times, the advent of cybernetics, electronic engineering, and computer science has furnished biologists with an even richer array of technological devices to employ as models in their explanations of the phenomena they investigate (cf. Grmek 1972; Vartanian 1973; Keller 1995; Canguilhem 2008; Reynolds 2018).

However, despite their undeniable heuristic value in certain experimental contexts, machine metaphors can be seriously misleading when they are used to ground the conceptualization of biological phenomena. The reason is that, some superficial similarities notwithstanding, living systems are fundamentally different from machines. Ontologically speaking, the *machine conception of the organism*, as I have referred to it in the past, is fraught with problems. In previous work (Nicholson 2013, 2014, 2018), I have advanced two major arguments against the theoretical understanding of organisms as machines, which still pervades many areas of contemporary biology. The first, which can be referred to as the Argument from Teleology, states that organisms are intrinsically purposive (in the sense that their activities and internal operations are directed toward the maintenance of their own organization), whereas machines are extrinsically purposive (given that their workings are geared toward fulfilling the functional ends of external agents). The second, which can be referred to as the Argument from Thermodynamics, states that organisms exhibit dynamic stability (due to their need to constantly exchange energy and matter with their surroundings to keep themselves in a negentropic steady state far from equilibrium), whereas machines exhibit static stability (given that they do not require to constantly expend free energy to ensure their continued preservation as they slide back and forth from equilibrium to near-equilibrium conditions). I have shown that the former argument has especially salient consequences for development and evolution (see Nicholson 2014), whereas the latter argument has particularly important implications for morphology, physiology, and bioenergetics (see Nicholson 2018).

In this chapter, I draw inspiration from a classic essay by Haldane titled *On Being the Right Size*, first published in 1926, to propose a third argument against the ontological identification of organisms as machines that is especially relevant for current research in cell and molecular biology. I shall refer to it as the Argument from Scale. Roughly, this states that, owing to their minuscule size, cells (and their macromolecular components even more so) are subject to very different physical conditions compared with much larger objects like machines. Machine metaphors and analogies draw on our intuitive familiarity with the macroscopic world of our everyday experience, but such intuitions fail us when we attempt to grasp the structure, function, and behavior of microscopic entities, as these exist in drastically different environments from our own (and our machines).

In the ensuing sections I will illustrate this argument by critically examining four core conceptual models of molecular biology that were originally imported from electronic and mechanical engineering (namely, "genetic program," "cellular circuitry," "molecular machine," and "molecular motor") and by showing that their explanatory deficiencies ultimately derive from their neglect of the impact of scale.[1] But first, it shall be useful to remind ourselves of the claims that Haldane put forward in his essay to better understand how they can be fruitfully redeployed in a contemporary context. At the end of the chapter, I will try to explain why so many molecular biologists today appear to have forgotten the importance of scale that Haldane famously drew attention to in his essay.

## Redeploying the argument of Haldane's
## *On Being the Right Size*

Although Haldane is primarily remembered for his foundational contributions to theoretical population genetics, he was also a prolific essayist and an avid popular science writer who wrote numerous articles on a variety of topics for a lay audience. *On Being the Right Size* is one of them. Its message is simple, but it has profound consequences. "For every type of animal," Haldane writes (1928, 20), "there is a most convenient size, and a large change in size inevitably carries with it a change of form." Size directly constrains the shape and structure that an animal can assume, as well as its behavior. An animal's way of life is conditioned by the physical forces that exert the greatest effect at the scale at which it exists.

For example, gravity poses no danger to a small animal, but it is a very serious threat to a large one.[2] As Haldane memorably puts it, "[y]ou can drop a mouse down a thousand-yard mine shaft; and, on arriving at the bottom, it gets a slight shock and walks away, provided that the ground is fairly soft. A rat is killed, a man is broken, a horse splashes" (ibid., 21). An insect is not afraid of gravity, as it has a negligible effect on its way of life; it can fall without danger and crawl up a wall or cling to a ceiling with remarkably little trouble. Conversely, surface tension is of little significance to a large animal but is of critical importance to a small one. A man coming out of a bath, Haldane observes, carries with him a film of water that is about half a millimeter thick and weighs about half a kilogram. A wet mouse, however, has to carry its own weight in water. And a wet fly has to lift many times its own weight. In fact, once a fly gets wet and falls in the grip of the surface tension of water, it is likely to remain there until it drowns. "An insect going for a drink is in as great danger as a man leaning out over a precipice in search of food" (ibid., 22), which is the reason why most insects keep well away from their drink by means of a long proboscis.

Many of Haldane's examples are based on the square-cube law, which states that the volume of a shape increases much faster than its surface area. Specifically, volume increases as the cube of length, while surface increases only as the square. A large organism has a far lower surface-to-volume ratio than a smaller organism of comparable shape. This explains why large animals have less trouble keeping warm than smaller animals, which cannot help dissipating more heat because of their higher surface-to-volume ratio. A mouse must eat about a quarter of its own weight in food every day just to keep warm. And although five thousand mice weigh as much as a man, their combined energetic consumption (through food and oxygen) is about seventeen times a man's. But small animals exploit their high surface-to-volume ratio in other ways. Insects have no need for complex circulatory systems; the oxygen their cells require can be directly absorbed by diffusion of air through invaginations in their external surface. In order to become larger, animals have had to evolve oxygen-carrying bloodstreams as well as pulmonary alveoli (to increase the surface area available for the exchange of gases) and a gastrointestinal tract (to increase the surface area available for the absorption of food). Haldane realized that "[t]he higher animals are not larger than the lower because they are more complicated. They are more complicated because they are larger" (ibid., 23). And the same can be said for plants. In general, "[c]omparative anatomy is largely the story of the struggle to increase surface in proportion to volume" (ibid.).

Such geometric relations determine the morphology and physiology of organisms and impose unbreachable limits on their possible dimensions. Given their existing morphology and physiology, it is simply not possible for insects to grow to be much larger than they already are. For the same reason, the giant creatures found in fantastical stories – from *Gulliver's Travels* to *Godzilla* – are impossible. Haldane calculated that a giant ten times as high as man, and also ten times as wide and ten times as thick, would weigh a thousand times more than man. But because the cross-sections of its bones would be only a hundred times those of man, every square centimeter of giant bone would need to support ten times the weight supported by every square centimeter of human bone. The consequence of this is that the giant would break its thighs every time it tried to take a step.

Of course, Haldane was not the first to reflect on the impact of size in biology.[3] Three centuries earlier, Galileo had made strikingly similar observations in his *Dialogues Concerning Two New Sciences*. Galileo already understood that size imposes fundamental constraints on the possible proportions of an organism, as well as of its parts. In his own words,

> An oak two hundred cubits high would not be able to sustain its own branches if they were distributed as in a tree of ordinary size; [similarly,] nature cannot produce a horse as large as twenty ordinary horses or a giant ten times taller than an ordinary man unless by miracle or by greatly altering the proportions of his limbs and especially of his bones, which would have to be considerably enlarged over the ordinary.
>
> (Galileo 1914 [1638], 4)

Galileo illustrated this by graphically depicting how the bone of a large animal must thicken *disproportionally* to provide the same relative strength as the corresponding bone of a small animal, as shown in Figure 2.1. As size increases, skeletal structure needs to become much stronger and more robust. This and many of Haldane's examples I have discussed above demonstrate the inevitability of allometric scaling. Animals are not isometric. Large animals do not look like small organisms scaled up in size, and vice versa. No one would mistake an elephant for a mouse, or a fly for an albatross, even if they are portrayed as being the same size. Quantitative changes in size necessarily entail qualitative changes in form and function. And this is as true for organisms as it is for other kinds of physical objects, such as ships, buildings, and *machines*.

Let us now see how these old insights can be put to work in the context of current cell and molecular biology. The most obvious feature of cells and molecules,

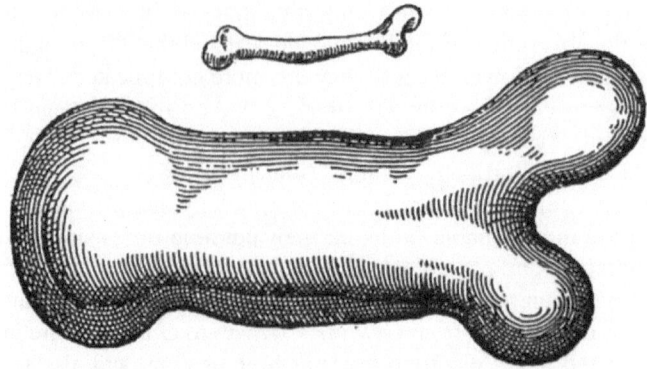

*Figure 2.1* Galileo's drawings showing the extent to which the shape and proportions of a bone would need to be modified for it to perform its function if its length was increased by a factor of three.

Source: Figure adapted from Galileo 1914 [1638]. Reproduced with permission.

especially when we compare ourselves to them, is that they are extraordinarily small. The difference in size between a man and a paramecium is several orders of magnitude greater than the one between a man and a giant, or between a mouse and an elephant. As a result, the morphological and physiological differences between them are also far more dramatic. I have already indicated that the structure, function, and behavior of every organism are adjusted to the scale at which it lives. This scale defines the physical environment in which each organism finds itself and determines the forces that have the greatest impact on its way of life. Cells and molecules are so miniscule that they exist in an environment that is *completely different* to our own. It is not an exaggeration to say that macroscopic and microscopic entities inhabit different "worlds." Whereas the macroscopic world is ruled by gravity and inertia, the microscopic world is governed by Brownian motion, which results from the thermal agitation of molecules above absolute zero. This has serious implications for the explanations we formulate of cellular and molecular phenomena. Most importantly, our imagination and intuition, based as they are on our experience of the macroscopic world, fail us when estimating the adaptive problems that cells and molecules have to overcome, as they inhabit a world that is utterly alien to us.

The main lesson that I wish to draw from this is that we should be extremely skeptical of analogies that seek to explain properties of microscopic entities by appealing to properties of macroscopic ones. Unfortunately, this is *precisely* what metaphorical appeals to machines in the modelling and explanation of molecular and cellular phenomena attempt to do. We routinely resort to machines to shed light on microscopic structures and processes because machines are familiar and intuitively intelligible macroscopic objects of our everyday experience. To conceptualize something as a machine is already to assume that we have a basic epistemic handle on how it works. The problem is that if our machines *were* the size of molecules they would not be able to function the way they do, as their physical environment would make it impossible for them to do so. It follows, therefore, that we should try to avoid distorting the reality of cells and molecules by construing it using concepts and models borrowed from the domains of electronic and mechanical engineering. Indeed, the reason why engineering metaphors in molecular biology mislead more than they illuminate, as the following four case studies will illustrate, is that they are not appropriately calibrated to the scale of the target domain they are called upon to explicate.

## Metaphor #1: "genetic program"

The first engineering metaphor I shall discuss is the perennially popular notion that every cell contains a *genetic program* that directs and controls its functions by executing a predetermined set of operations according to instructions encoded in its genes. This idea was proposed, seemingly independently, by Jacob and Monod (1961) and by Mayr (1961), and it quickly garnered widespread acceptance among molecular biologists.[4] Jacob (1973, 9) admitted that the genetic program "is a model borrowed from electronic computers. It equates the genetic material of the egg with the magnetic tape of a computer," adding that "everything urges one to compare the logic of heredity to that of a computer. Rarely has a model [. . .] proved to be more faithful" (ibid., 265). Mayr (1982, 106), for his part, remarked that "all manifestations of development and life are controlled by genetic programs," noting that "[n]othing comparable to it exists in the inanimate world, except for manmade computers" (ibid., 55). More than half a century after it was first proposed, the metaphor continues to pervade the specialist as well as the popular scientific literature (e.g., Danchin 2009; Bray 2009).

Despite its enduring popularity, the problems with the genetic program have long been pointed out by biologists and philosophers (e.g., Webster and Goodwin 1982; Atlan and Koppel 1990; Nijhout 1990; Moss 1992; Keller 2000; Oyama 2000; Longo and Tendero 2007; Nicholson 2014). It has been repeatedly argued, for example, that the metaphor is conceptually incoherent, as the genetic program requires its own output to be executed: the protein "hardware" that runs the genetic "software" is not independent of it but is itself produced by that very software. What has received far less attention, however, is the fact that the difficulties involved in theoretically transferring the idea of a program governing the operation of a computer to the way genes are involved in specifying the cellular phenotype have a great deal to do with the physical scale at which the latter process takes place. It is because of this, for instance, that the deterministic assumptions of the genetic program model of development are completely unrealistic. It is just not possible, physically speaking, to "compute the embryo" from a complete description of a fertilized cell's DNA sequence and the location of all its proteins (cf. Wolpert 1994; Rosenberg 1997). One reason for this is very simple.

Gene expression is first and foremost a *molecular* process, and like all molecular processes it is subject to the dampening stochastic effects of Brownian motion. Let me elaborate this point a little.

Gene expression is an extremely intricate process. Consider how it gets started: an inducer, which can be an intracellular or extracellular signal, triggers a chain of biochemical reactions that causes proteins called activators to bind to specific sites in the DNA known as enhancers. Upon binding, the activators interact with other proteins that recruit RNA polymerase and its associated transcription factors to the promoter region of the target gene, where it begins the process of transcription. Numerous additional steps need to be strictly followed after transcription, including RNA processing and export, translation, and protein folding and sorting (Alberts et al. 2008). The point is that for even a single protein to be successfully expressed in the cell, a huge number of molecules need to interact with one another in exactly the right way, at exactly the right time, and in exactly the right order. And it should not be surprising that the likelihood that all of this happens in a perfectly efficient and precisely timed fashion (as one would expect of the programmatic execution of an algorithmic sequence of coded instructions) is virtually zero once we take into account the random and ferocious buffeting that all molecules are subject to inside the cell by virtue of their size.

The impact of stochasticity on gene expression is exacerbated even further by the fact that a cell, unlike a test tube, contains very low copy numbers of the relevant molecules. There are only one or two copies of any given gene in a cell, just a few copies of each mRNA molecule, and a few dozen copies of the required polymerases and transcription factors (Xie et al. 2008). Consequently, it is not possible to appeal to the law of large numbers to make accurate predictions about the process. Of course, it is still possible to make predictions when gene expression is measured across a *population* of cells, as the individual differences between cells are averaged out, but this becomes impossible when measuring the expression of a gene in a single cell. The recent introduction of methods capable of tracking individual molecular reactions on a cell-by-cell basis has confirmed that even genetically identical cells subject to the same external conditions exhibit substantial variability in their gene expression profiles due to the inherent stochasticity of the process (Altschuler and Wu 2010). This finding makes perfect sense given the scale at which gene expression occurs, but it is difficult to reconcile with the genetic program model, as two identical computers running the same software program are expected to execute it in exactly the same way.

Frustratingly, instead of questioning the theoretical adequacy of the genetic program (or the assumptions that underlie it), molecular biologists initially reacted to the discovery of the stochastic character of gene expression by borrowing an additional idea from the realm of engineering to make sense of it; namely, the concept of *noise* (e.g., Elowitz et al. 2002; Rao et al. 2002; Raser and O'Shea 2005). In engineering, noise refers to an undesirable random disturbance that garbles the transmission of a message. Noise is therefore a nuisance that engineers strive to overcome by designing machines that filter out its detrimental effects. The rationale for appropriating this term seems to have been that stochasticity thwarts the capacity of molecular biologists to perfectly predict cellular behavior in the same way that noise thwarts the capacity of engineers to design and manufacture totally predictable machines. In any event, the analogy does not hold because cells, unlike machines, actually benefit from the "noisiness" of gene expression. Far from being disruptive or detrimental, recent research has shown that gene expression noise plays many critical biological functions. In microbial cells it is a key generator of phenotypic diversity within populations and therefore serves to increase their adaptability to new environmental conditions. And in eukaryotic cells, among other things, it helps to determine cell fate decisions, thereby shaping the way in which cells differentiate during development (Eldar and Elowitz 2010; Balázsi et al. 2011).

Overall, the genetic program can only be understood as a rather crude approximation of what happens during gene expression. Although the metaphor does compellingly capture the order, reliability, and robustness of this process, it does so at the expense of abstracting away the messy molecular details that enable it to take place. Upon close inspection, the analogy with how a computer executes a program breaks down. Lewontin (2000, 17), as ever, puts it best when he declares that "[a]ny computer that did as poor a job of computation as an organism does from its genetic 'program' would be immediately thrown into the trash and its manufacturer would be sued by the purchaser." It is worth pointing out as well that the genetic program metaphor misrepresents not just the phenomenon it seeks to explain but also the way in which scientists investigate it: molecular biologists are simply not in the business of deducing computable functions, either mathematical or algorithmic, from their empirical studies.[5]

## Metaphor #2: "cellular circuitry"

Another engineering metaphor commonly used by molecular biologists, and which is to some extent implied by the genetic program, is the notion of *cellular circuitry*. This is the idea that the programmatic instructions encoded in the genes are carried out in a logical fashion by fixed, solid-state circuits inside the cell that mimic the circuit boards of electronic engineering. The metaphor of cellular circuitry goes beyond that of the genetic program because it argues that computers are not merely functionally analogous to cells and other biological systems but also *structurally* analogous. In other words, not only does a cell behave in a programmed way, but its internal architecture also displays the modular organization that is typical of the hardware of an electronic computer.

There are two main areas of current research in which the cellular circuitry metaphor is regularly employed. The first is in relation to gene regulation, particularly as it pertains to embryonic development. Here the metaphor provides the conceptual foundation for the understanding of gene regulatory networks (GRNs). GRNs are comprised of cis-regulatory elements (i.e., the regions in the vicinity of each gene that contain the specific sequence motifs at which the regulatory proteins that affect its expression bind) plus the set of genes that encode these specific regulatory proteins. Conceptualized through the lens of electronic engineering, GRNs are characterized as hierarchical assemblies of "modular subcircuits and their interconnections" (Davidson 2009, 535), where each sub-circuit is an "information processing unit" (Davidson 2001, 7) that produces a discrete developmental output, which is defined in terms of the effect it has on the spatial or temporal expression pattern of a particular gene. Importantly, these outputs can be mathematically represented as combinations of Boolean operators (e.g., AND, OR, NOT), so that the entire GRN can be viewed as "a logic process-ing system" made up of distinct "computational devices, the functions of which are conditional on their inputs" (Davidson and Levine 2005, 4935). The GRN for the early development of the sea urchin embryo is shown in Figure 2.2, which illustrates just how complex and detailed these models have become, and also just how uncanny, and intentional, their resemblance is to the wiring diagrams of electronic engineering.

The first thing to bear in mind when evaluating GRN circuits is that they include only a fraction of the genes, cis-regulatory elements, and proteins involved in the developmental process. Moreover, despite their seemingly robust design, the depicted circuits have rather restricted predictive capabilities, as their computing power is dependent on the presence of very specific environmental conditions. As GRN researchers readily admit, "[w]e do not know how they [i.e., GRNs] would

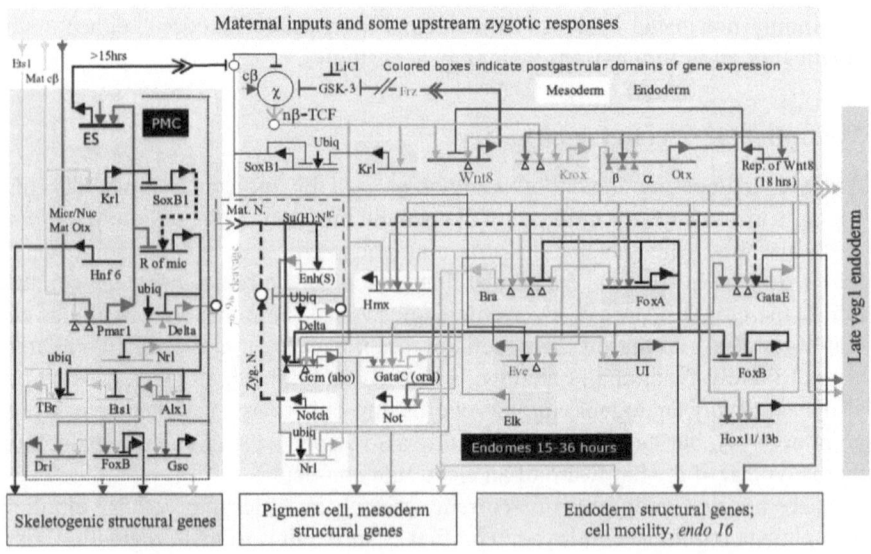

*Figure 2.2* Gene regulatory network for the early development of the sea urchin embryo. The circuit is divided according to the embryonic region in which each gene is expressed. The lines with bent arrows represent the transcription pattern of the genes named beneath them, as inferred from experimental studies.

Source: Figure adapted from Davidson et al. 2003. Reproduced with permission.

Note: This figure can be accessed in color via the eBook version of the book and eResources at www. routledge.com/9780815380788.

behave even in a slightly different context (both abiotic and biotic)" (Wang and Buck 2012, 382). Still, the most serious problem with GRN models is that they misinterpret our ability to describe certain patterns of transcriptional activity in terms of Boolean operators as empirical proof that cis-regulatory elements in the genome and their associated regulatory proteins causally interact in a perfectly reproducible, deterministic manner. It is a blatant – even if often convenient – idealization to characterize molecular processes such as transcription and differentiation in terms of computable logic functions, and the reason, as I have already discussed, has to do with the scale at which they take place. Every single step in these processes (as with every biochemical reaction in the cell) relies upon probabilistic collision events between small numbers of randomly moving molecules, and these stochastic effects are amplified in regulatory cascades. This imposes absolute limits on the predictive capabilities of these models, and it is also why the analogy with electronic circuit boards is inappropriate and frequently misleading.

The second context in which the cellular circuitry metaphor is widely used is in the study of protein-protein interactions, particularly signal transduction pathways, which enable cells to make decisions, such as whether to grow, differentiate, move, or die. "The analogy between cell signaling and man-made machines," Mayer et al. (2009, 81.1) observe, "is all-pervasive, frequently adopting the imagery of [. . .] electronic circuit boards." The reason, according to Dueber et al. (2004, 690), is that signal transduction pathways "have information-processing capabilities that rival computers: they can perform complex signal integration [and] switch states in a manner that retains memory or generate complex temporal behaviors, such as oscillations." They are also presumed to be analogous in their organization: "[j]ust as electronic circuits are built of simpler components cellular signalling circuits are composed from a modular toolkit of components" (ibid.). Specifically, "transistors are replaced by proteins (e.g., kinases and phosphatases) and the electrons by phosphates and lipids" (Hanahan and Weinberg 2000, 59). Figure 2.3 shows a typical example of how these pathways are represented in the literature.

It is hard to resist the appeal of diagrams of this kind. Besides economically summarizing a wealth of information about how particular proteins interact, by deliberately imitating the design charts of electronic circuits, with their neat modular structure and their reassuring arrows, these attractive representations convey the comforting impression of understanding and control. However, in order for them to be as explanatorily useful as the diagrams of engineering, they must assume a very high degree of specificity in the molecular interactions that are depicted as arrows. The trouble is that this assumption is not well supported empirically. A growing body of experimental evidence suggests that exquisite specificity in protein function is the exception rather than the rule (Nobeli et al. 2009; Kupiec 2010).[6] What a protein does in the cell is determined as much by the milieu it finds itself in as by its amino acid sequence. The same polypeptide chain can partake in a wide variety of cellular functions depending on where and when it is expressed; a rather unexpected phenomenon that has been dubbed "moonlighting" (Jeffery 2003; Copley 2003). Moonlighting occurs because proteins *in*

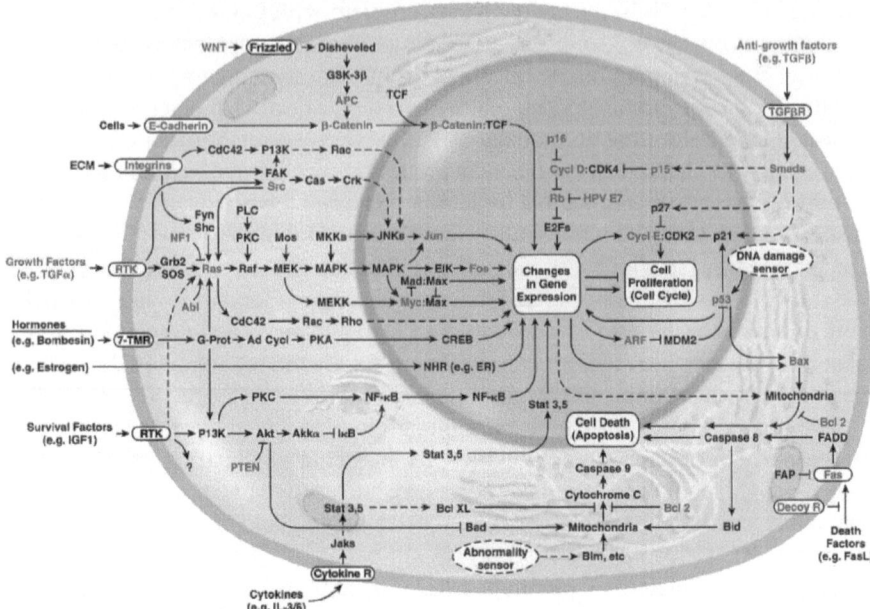

*Figure 2.3* Signal transduction pathways represented as wiring diagrams to reflect what Hanahan and Weinberg (2000: 59) call the "integrated circuit of the cell."

Source: Figure adapted from Hanahan and Weinberg 2000. Reproduced with permission.

Note: This figure can be accessed in color via the eBook version of the book and eResources at www. routledge.com/9780815380788.

*vivo* actually interact with many more binding partners than was previously supposed (Gierasch and Gershenson 2009). Nevertheless, reports of moonlighting become far less surprising when we bear in mind that most proteins are constantly colliding with one another as a result of being violently knocked about by Brownian forces. This is all a consequence of the strange, stochastic world that proteins inhabit by virtue of being so small.

The fact that every protein in the cell can potentially associate with a large number of other proteins leads to a dazzling explosion of combinatorial possibilities that is exceedingly difficult to faithfully represent in diagrammatic form. The problem with circuit-like characterizations and representations is that signal transduction pathways do not exist as discrete, mutually exclusive subcellular compartments, given that the proteins that constitute them participate in many other pathways, as well as in other, altogether different cellular processes. Signaling cascades in the cell are deeply interconnected; they interact, or "cross-talk," with one another in numerous ways (Knight and Knight 2001). Even the most straightforward textbook representations of linear sequences of protein-protein interactions tend to mislead, as "the simple causal links that are being depicted hide an underlying complexity that is often essential to explain real world functionality: so much is swept under the rug" (Blinov and Moraru 2012, 3). A dramatic illustration of this was provided by Dumont and colleagues, who, in an attempt to diagrammatically represent experimentally verified cross-signalings between four distinct cascades (as reported in the literature during the previous two years), produced a remarkable, yet utterly unreadable, "horror graph," shown in Figure 2.4, in which, according to the authors, "everything does everything to everything" (Dumont et al. 2001, 457).

When visualizing cellular circuit diagrams, it is important to understand that they represent only one of the many potential ways in which a given set of proteins can interact with one another. Tweak the intracellular or extracellular context ever so slightly and the wiring between the proteins will change. And, of course, we should not forget that there is no actual wiring physically connecting proteins as there is in a real electronic circuit. Instead – and this is, again, a consequence of their size – proteins exist in a fluid and dynamic environment in which they rely on probabilistic collision events with appropriate partners to reliably perform particular cellular functions at particular times. Diagrams such as Figure 2.3 wrongly imply that the proteins featured in them always form the same exact networks of interactions, which are envisaged as

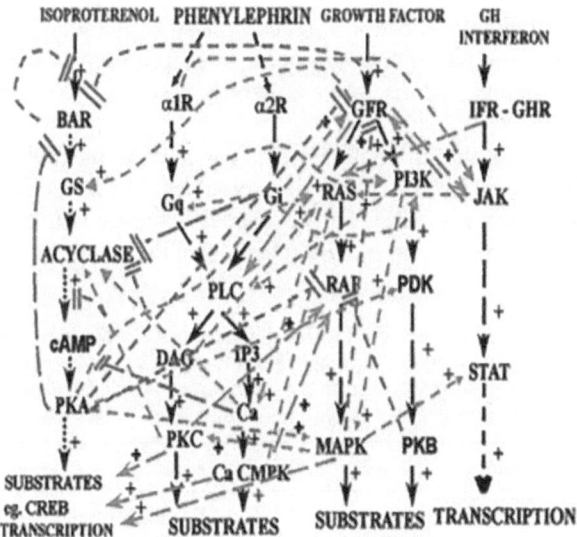

*Figure 2.4* "Horror graph" indicating cross-signalings between four signal transduction cascades. The black arrows show the "textbook" representation of the cascades. The remaining arrows denote all the experimentally verified cross-signalings reported in the literature in the space of only two years. Despite featuring only four cascades with four to six steps each, the total number of possible interaction combinations is 760!

Source: Figure adapted from Dumont et al. 2001. Reproduced with permission.

Note: This figure can be accessed in color via the eBook version of the book and eResources at www. routledge.com/9780815380788.

fixed, solid-state circuit boards. In doing so, these diagrams prevent us from appreciating the vast spectrum of alternative interaction networks that the same set of proteins can and do form in different cells, and even in the same cell at different times. The majority of protein-protein interactions are contingent and opportunistic (Misteli 2001; Kurakin 2009). There is no predetermined "design" that specifies the way in which the proteins in a cell interact any more than there is a program of genetic instructions that is deterministically computed by the cell (or the embryo). Conceptualizing protein-protein interactions as circuits may seem like a harmless heuristic simplification, but it can mislead us into thinking that we understand more than we actually do. Worse still, it can inadvertently direct our attention away from the factors and causal relations that may turn out to be most relevant for explaining the phenomena we are interested in.

## Metaphor #3: "molecular machine"

A further engineering metaphor that has completely permeated the molecular biology discourse is the concept of *molecular machine*. This notion, which borrows more from mechanical than from electronic engineering, has become central to the way protein complexes and many other subcellular assemblies are conceptualized (Block 1997; Piccolino 2000). The success of this metaphor lies in its versatility. As Table 2.1 illustrates, an extremely wide range of machines can be summoned on its behalf to give substance to descriptions of the structure and function of macromolecular assemblies, thereby rendering them more tractable and familiar.

But what exactly is the rationale for using the term "machine" to designate macromolecular assemblies? According to Nogales and Grigorieff (2001, F1), "this designation captures many of the aspects characterizing these biological complexes: modularity, complexity, cyclic function, and, in most cases, the

*Table 2.1* Examples of the different kinds of machines that molecular biologists draw upon as conceptual resources to ground their characterizations of macromolecular assemblies in the cell.

| Macromolecular assembly | Machine |
| --- | --- |
| Cilium, flagellum | Propeller |
| ATP synthase | Generator |
| Ribosome | Factory assembly line |
| Ion channel, nuclear pore | Gate, key, pass |
| Polymerase | Copy machine |
| Ligase | Chain coupler |
| Spliceosome | Film editing machine |
| Protein targeting mechanism | Mail sorting machine |
| Proteasome, apoptosome | Bulldozer, destroyer |
| Magnetosome | Compass |

consumption of energy." Frank (2011, 1), for his part, argues that " '[m]achine' is useful as a concept because the molecular assemblies [. . .] share important properties with their macroscopic counterparts, such as processivity, localized interactions, and the fact that they perform work toward making a defined product." And Browne and Feringa (2006, 26), when asking "What makes a molecule a machine?", answer that in "a molecular machine we are able to switch between two [or more] molecular states (shapes) in a controlled manner as part of a repetitious mechanical cycle." Finally, Alberts, who has been one of the most influential advocates of the molecular machine metaphor in molecular biology during the last two decades, gives the following explanation:

> Why do we call the large protein assemblies that underlie cell function protein *machines*? Precisely because, like the machines invented by humans to deal efficiently with the macroscopic world, these protein assemblies contain highly coordinated moving parts. Within each protein assembly, intermolecular collisions are not only restricted to a small set of possibilities, but reaction C depends on reaction B, which in turn depends on reaction A – just as it would in a machine of our common experience.
>
> (Alberts 1998, 291)

The idea, then, is that macromolecular assemblies in the cell can be legitimately thought of as machines because they effectively capture in their operation the high degree of coordination and precision that is typical of mechanical devices. However, the scale at which these macromolecular assemblies operate makes this comparison hard to uphold from a physical point of view. Perfectly orchestrated mechanical movements are simply not possible in a world that is governed by Brownian motion. Even the structure of a macromolecule cannot be compared to that of a machine. Machines tend to rely on a hard and rigid constitution for their operation. Proteins, on the other hand, exhibit very high degrees of structural flexibility. In fact, it is becoming apparent that, in their native environments, proteins behave more like liquids than solids; they can be characterized as "dense liquids" or "melted solids," consisting of a "near-solid interior" and a "full-liquid exterior" (Rueda et al. 2007, 798; see also Zhou et al. 1999). What is more, it is now widely acknowledged that most proteins do not have a single ordered conformation. What we refer to as *the* conformation of a protein actually comprises an entire spectrum of well-defined configurations separated by low-energy barriers that the protein continuously samples by means of stochastic fluctuations (Yang et al. 2003). Even more counterintuitive is the discovery that many proteins do not have an ordered conformation *at all*. These have come to be known as "intrinsically disordered proteins" (Uversky 2013; Wright and Dyson 2015) and they empirically refute the longstanding mechanical assumption that a protein needs to have a clearly defined three-dimensional structure for it to perform its function; a requirement that is, of course, crucial for the operation of a mechanical device. Macromolecular assemblies, which are primarily composed of protein subunits, therefore lack most of the structural characteristics that we associate with machines, often exhibiting instead fluid, ever-flickering, "fuzzy" structures (Fuxreiter and Tompa 2012).[7]

The fundamental shortcomings of the molecular machine metaphor have far-reaching consequences for how macromolecular assemblies in the cell are studied and represented. For example, they call into question the adequacy and usefulness of virtual movies that purport to faithfully depict how these assemblies modify their structure as a result of their operation. Such movies are made by using cryo-electron microscopy to visualize a frozen population of isogenic macromolecules in a near-native state, categorizing each macromolecule in the snapshot according to its reconstructed three-dimensional structure, and ordering these static reconstructions so as to create the impression of motion. "Morphing" computer software is then used to interpolate additional hypothetical frames to smoothen the transition between reconstructions and prevent the resulting movements from appearing excessively jerky (Moore 2012; Nogales 2016). Thus, unlike conventional live imaging microscopy techniques in which what one sees more or less reflects what is really happening, in molecular movies the temporal dimension is introduced *virtually* by linking unrelated reconstructions of different macromolecules to plausibly infer a coherent "time line" of a single macromolecule.

An important limitation of these movies is that it is not possible to conclusively determine whether the conformational trajectories devised by morphing programs are accurate, let alone that such trajectories are always followed by every macromolecule of the type depicted in the movies. The problem is that, because macromolecules are so often thought of as molecular machines, molecular biologists tend to assume, incorrectly, that they move in a mechanical fashion. This has been forcefully pointed out by Moore (2012) with regard to movies of the ribosome – that most paradigmatic of molecular machines (Garrett 1999; Frank 2000). Moore argues that a virtual movie makes the ribosome appear to be something it is not:

> Like the structures on which it is based, the movie will actively invite viewers to think that the ribosome works the same way as a clock, or a machine for making candy bars. *It is no help that macromolecules* [. . .] *are commonly called molecular machines.* The use of the word 'machine' in this context is pernicious because of its implication that the functional properties of macromolecules can be explained mechanically, *which is simply not true.*
>
> (Moore 2012, 7–8; emphasis added)

Due to their minuscule size, ribosomes (and smaller macromolecules even more so) cannot possibly operate in the orderly and reproducible manner that is characteristic of machines. In a machine, as we noted earlier, the motions of the various parts are perfectly coordinated. For example, when a gear rotates, the shaft to which it is connected rotates in synchrony, a spring is compressed, a latch is released, etc. All of these movements are purposeful and predictable and are always precisely executed in exactly the same temporal sequence. Macromolecular assemblies, by contrast, are subject to continuous Brownian motion, which means that the vast majority of conformational changes they undergo are the result of "random walks" that have nothing to do with their function.[8] This is very significant because if the usefulness of a virtual movie is predicated on its ability to explain the function of a macromolecule on the basis of its conformational changes, then it follows that a *perfectly accurate* movie (i.e., one that realistically depicted all of the macromolecule's random motions) would be of no explanatory value whatsoever. Virtual movies of mechanically moving macromolecules are undoubtedly fun to watch, but they are also misleading – especially when shown to impressionable students or to the unsuspecting general public. As Moore (2012, 15) himself concludes, "[s]tructure-based movies of ribosome function should have a surgeon-general's warning attached to them because they are more likely to deceive the unwary than enlighten them."

## Metaphor #4: "molecular motor"

The final engineering metaphor I shall examine is the concept of *molecular motor*, which is used to characterize proteins responsible for transporting cargo to specific destinations inside the cell. Although it is generally regarded as a subclass of the more general notion of molecular machine, its usage poses its own set of distinct challenges that merit separate attention. For a start, it could be argued that the concept of motor does not necessarily imply the concept of machine. If we understood a motor simply as an entity that imparts motion – which is actually the first definition of "motor" listed in the *Oxford English Dictionary* – then there would be nothing metaphorical about referring to kinesin, dynein, and myosin as motors. In practice, however, the designation "molecular motor" in molecular biology tends to carry clear connotations of machines and of mechanical engineering. When proteins capable of directional movement are described as molecular motors in the literature, what is typically implied is that that they resemble macroscopic mechanical motors with regard to their structure and to their operation. In fact, it is not unusual for them to be compared to automobiles. Both, it is argued, consume fuel to power their motion. Moreover, Vale and Milligan remark that:

> Just as in an automobile, the site that processes the chemical fuel [in a molecular motor] must be linked through intermediate components to the site that ultimately generates the motion. In the automobile, the breakdown of the chemical fuel is coupled to the stroking of a piston, which in turn is linked through the crankshaft and transmission to the turning of the wheels. A somewhat analogous situation for translating chemical changes into mechanical motions exists in molecular motors.
>
> (Vale and Milligan 2000, 90)

Specifically, the claim is that the energy released from the chemical fuel is used to induce a large-amplitude conformational change in the motor protein, which generates a mechanical force – a "power-stroke" – that drives the molecule forward relative to a polymeric track (Howard 2001; Tyska and Warshaw 2002). Sometimes, this power-stroke is compared to the mechanical release of a viscoelastic spring (e.g., Howard 2006). In the case of kinesin, which has dimeric "legs" that alternatively attach to tubulin, the repetitive power-strokes result in directed movement that makes the protein appear like a tiny robot walking along the microtubule, and this is indeed the way in which its motion is usually represented in diagrams and animations (e.g., Asbury 2005).

Once again, the problem with these familiar mechanical models inspired by our everyday experience of the macroscopic world is that they fail to recognize the drastically different physical conditions that characterize the microscopic world. When we are walking, the two major physical forces at play are gravity and inertia. Most of the motive power is expended by repeated cycles of acceleration, as the foot that was in touch with the ground is brought forward to a position in front of the torso. Friction plays only a minor role as far as the energetics are concerned. In the microscopic world, however, the impact of inertia (which is proportional to volume and mass) is completely dwarfed by the impact of friction (which is proportional to surface area). The high viscous friction (or drag) of water at the molecular scale means that, for a bacterium, swimming in water feels like what swimming in molasses would feel to us (Bier 2003; Astumian 2007). Moreover, although we might (just about) be able to imagine what it would feel like to be immersed in molasses, it is much harder to imagine another feature of aqueous solutions that cells and their macromolecular components experience by virtue of being so small: the molasses that surrounds them is furiously moving about as a result of the thermal agitation of the water molecules. We have to remember that a motor protein does not experience water as a fluid continuum in the way that we do, but as an extremely dense array of rapidly moving particles that are constantly striking it from all sides. "Even a freak hailstorm," Astumian (2001, 58) writes, "does not come close to the tempestuous bombardment that is routine in the molecular world, but the effects can be analogous."

From a physical perspective it is difficult to understand how a motor protein could possibly walk in a directed manner by means of mechanical cycles of precisely coordinated power-strokes once we realize that "[f]or molecules, moving deterministically is like trying to walk in a hurricane: the forces propelling a particle along the desired path are puny in comparison to the random forces exerted by the environment" (ibid., 57). Recently, a growing number of researchers have come to appreciate that if we are to understand the way in which motor proteins move, we need to "[i]magine living in a world where a Richter 9 earthquake raged continuously" (Oster and Wang 2003, 207). So how exactly do motor proteins manage to move directionally in such a turbulent and chaotic environment? There are two alternatives: motor proteins must either work with the raging Brownian storm that engulfs them or fight against it, and in light of the above considerations, the former appears to be the preferable option. This has led to the hypothesis that motor proteins are not mechanical motors but *Brownian motors*. Instead of moving directionally by generating a large mechanical force that overpowers the stochastic effects of Brownian motion, motor proteins are thought to move by biasing the existing Brownian motion in a particular direction.

Figure 2.5 illustrates how Brownian motors harness stochasticity to move directionally. According to this model, motor proteins use the energy released from the chemical fuel they consume to switch between two alternative conformational states – "on" and "off" – with different energy profiles. When the motor proteins are on, their energy landscape has a jagged, sawtooth shape, and consequently random collisions jostle them overwhelmingly to the right, where they get trapped in the nearest energy minima. When the motor proteins are off, their energy landscape has a flat shape, and consequently random collisions cause them to perform random walks, with equal probabilities of moving to the left or to the right of their initial position. Thus, by periodically switching between on and off states through the repeated consumption of chemical energy, and by taking advantage of the incessant Brownian motion that characterizes their environment, motor proteins are able to move directionally in the absence of mechanical forces (Ait-Haddou and Herzog 2003).

When trying to comprehend how a motor protein moves along a cytoskeletal track, the assumption has long been that at least some of the mechanical principles "that have been derived by the engineers who analyse the machines of our common experience are likely to be relevant" (Alberts 1998, 291). But if the Brownian motor model of intracellular transport is even partially correct, then this attitude is bound to lead researchers astray. Due to their huge disparity in size, mechanical motors and Brownian motors operate according to fundamentally different principles. The former use energy to drive motion, whereas the latter use energy to restrain it. The former move despite stochastic fluctuations; the latter move because of them. The structure of the former must be hard and rigid, while that of the latter can be soft and plastic. In addition, Brownian motors are far more efficient than mechanical motors because they convert chemical energy directly into work without using heat or electrical energy as intermediates. An important upshot of

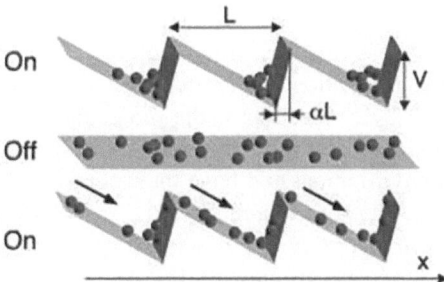

*Figure 2.5* Directed movement by Brownian motors. During the on phase, Brownian motors (shown as particles) move toward the closest energy trough. During the off phase, they undergo one-dimensional isotropic diffusion. Stochastically alternating between the two states results in net movement along the x axis.

Source: Figure adapted from Linke et al. 2005. Reproduced with permission.

Note: This figure can be accessed in color via the eBook version of the book and eResources at www.routledge.com/9780815380788.

relying on stochasticity for their operation is that the directional movements of Brownian motors are purely *statistical* occurrences. The timing of their individual journeys, as well as their precise trajectories, is non-deterministic and therefore impossible to predict. Every Brownian motor performs a unique "dance" despite moving in the same general direction. Overall, it is clear that attempts to draw on the properties of macroscopic motors to shed light on the properties of microscopic ones (such as the motor proteins inside the cell) are more likely to cloud and obfuscate than they are to clarify and illuminate.

## How did molecular biology come to neglect the impact of scale?

Before bringing this chapter to a close, it is worth pausing for a moment to consider the intriguing historical puzzle that the examination I have provided suggests. A growing number of molecular biologists are beginning to question the value of using metaphors and models imported from electronic and mechanical engineering, and this undoubtedly reflects an increasing awareness of the importance of adjusting explanations of molecular and cellular phenomena to the scale in which they take place. The odd thing about this is that it has taken molecular biologists so long to start taking the importance of scale seriously. Haldane was not a lone voice when he drew attention to the importance of size in his essay of 1926. The impact of scale was widely recognized at the time, remaining an important consideration in biological discussions during the first half of the 20th century. Take, for example, the second chapter of the revised edition of Thompson's celebrated magnum opus, *On Growth and Form*. It is titled "On Magnitude" and it presents a wonderfully detailed analysis of the numerous ways in which physical forces at various scales affect the lives of organisms of different sizes. In fact, its final paragraph eloquently articulates the basic thesis I have sought to defend in this chapter:

> [The world of] Man is ruled by gravitation. [. . .] [But in the] world where the bacillus lives, gravitation is forgotten, and the viscosity of the liquid, the resistance defined by Stokes's law, the molecular shocks of the Brownian movement, doubtless also the electric charges of the ionized medium, make up the physical environment and have their potent and immediate influence on the organism. *The predominant factors are no longer those of our scale; we have come to the edge of a world of which we have no experience, and where all our preconceptions must be recast.*
>
> (Thompson 1942, 77; emphasis added)

These remarks were written over three quarters of a century ago, so why do they now seem more relevant than ever? Or, to put it slightly differently, how did we come to forget what we used to know? It is obviously not possible to do justice to such a complex question here. In what follows I only wish to propose and briefly discuss a factor that might have contributed to molecular biology's neglect of scale during the second half of the 20th century; namely, the influence of Schrödinger's argument regarding the stability of the gene laid out in his famous little book *What is Life?* published in 1944.

Like all science classics, *What is Life?* is far more often cited than read. But if one bothers to go back and actually read how Schrödinger arrives at his well-known characterization of genes as "aperiodic crystals," the striking thing about his argument is that it is based primarily on considerations of size and scale! Schrödinger begins his book by noting that atoms, as a consequence of being so small, are incapable of exhibiting orderly behavior on their own because they are continuously subject to the stochastic effects of thermal agitation at any temperature above absolute zero. This is why physical laws are statistical in nature. Order and regularity can only emerge upon consideration of enormous numbers of atoms (or molecules), which collectively display macroscopic patterns of order. Schrödinger calls this the "order-from-disorder" principle, and he discusses several physical examples to illustrate it.

One of them, shown in Figure 2.6, concerns what happens when you fill a glass vessel with fog consisting of minute droplets. Over time, the fog gradually

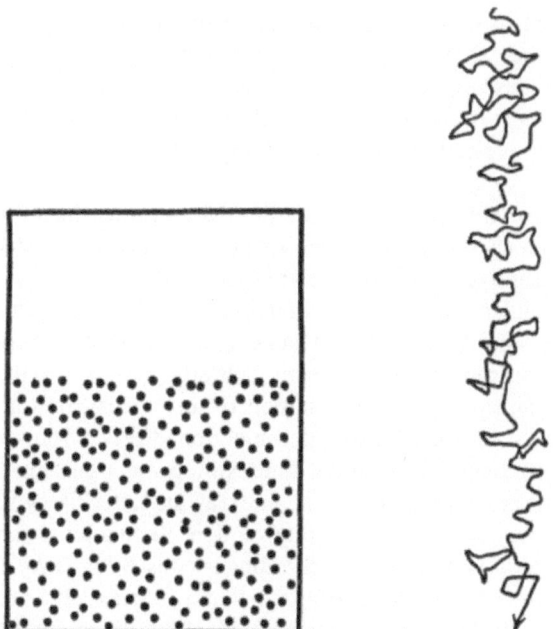

*Figure 2.6* One of Schrödinger's own illustrations of the "order-from-disorder" principle. The vessel on the left shows the regular, orderly sinking of fog over time. The arrow on the right delineates the irregular and disorderly trajectory of an individual droplet. The law-like behavior of the fog reflects a statistical average of the collective behavior of all the droplets of which it is composed.

Source: Figure adapted from Schrödinger 1944. Reproduced with permission.

sinks to the bottom with a well-defined velocity, determined by the viscosity of the air and the size and specific gravity of the droplets. Still, if you observe one of the droplets under the microscope you find that it does not permanently sink with constant velocity, but instead performs highly irregular movements – Brownian motion – as a consequence of thermal agitation. So, although the behavior of any given droplet is stochastic and disorderly as it sinks, the behavior of the fog *as a whole* is regular and orderly. In general, the larger the number of participating particles in a physical process, the more accurate the lawful prediction of its behavior. This is what mathematicians refer to as the law of large numbers.

Now, a "naïve physicist," Schrödinger writes, might be forgiven for thinking it self-evident that the astounding regularity and orderliness displayed by an organism must also be based on the macroscopic law-like patterns of behavior exhibited by large ensembles of interacting molecules. However, Schrödinger continues, "this expectation, far from being trivial, is wrong" (Schrödinger 1944, 20). His reasoning is as follows. The order of an organism is essentially determined by its genes, and we know from experimental studies that a gene molecule is not much larger than a few thousand atoms. This number, Schrödinger observes, "is much too small (from the [law of large numbers] point of view) to entail an orderly and lawful behavior according to statistical physics" (ibid., 21). Because genes are so tiny, they should not be able to reliably code for heritable traits, given that they are firmly in the grip of thermal agitation. And yet we know for a fact that genes *are* remarkably stable, "with a durability or permanence that borders upon the miraculous" (ibid., 49).[9] So how do we reconcile the small size of genes with their extraordinary stability in the face of constant stochastic perturbations?

Schrödinger's answer is that the genetic material must have the rigid, solid-state structure of a crystal, as only then would it be able to effectively withstand the relentless disruptive effects of Brownian motion. But unlike normal crystals, which display regular and periodic configurations, the structure of the genetic material must be "aperiodic" so that it can contain within it the "code-script" that specifies the organization of the organism. Schrödinger refers to the kind of order displayed by organisms as a manifestation of an "order-from-order" principle, which he explicitly contrasts to the aforementioned order-from-disorder principle described by statistical mechanics. Interestingly, he argues that in this crucial respect organisms are analogous to machines, as the latter likewise exhibit rigid, solid-state structures capable of resisting random fluctuations, enabling them to operate in an orderly way. Indeed, Schrödinger ends *What is Life?* by declaring that "the clue to the understanding of life is that it is based on a pure mechanism, a 'clock-work' [. . .] [that] also hinges upon a solid – the aperiodic crystal forming the hereditary substance, *largely withdrawn from the disorder of heat motion*" (ibid., 82, 85; emphasis added).

Schrödinger's deliberations led him to conclude that the solid-state, crystal structure of the genetic material renders it impervious to the physical forces that exert the greatest effect at the microscopic scale. Genes behave as if they were at absolute zero, as they do not appear to be affected by thermal agitation. And although he does not go into detail, his argument implies that the order encoded in the aperiodic crystal must somehow be reliably transmitted to the rest of the cell's components, especially to the proteins, so that these can individually express it through their functions in a way that similarly eludes or overcomes the raging Brownian storm of the molecular realm.

What I want to suggest here is that the influence of this idea – which is central to Schrödinger's argument in *What is Life?* – is responsible, at least in part, for molecular biology's subsequent neglect of the importance of scale. Just as Schrödinger had done with genes, molecular biologists went on to focus on the structure of proteins and other macromolecular assemblies (using methods such as X-ray crystallography), drawing attention to their crystal-like stability and rigidity, emphasizing their functional specificity, and ignoring the chaotic, destabilizing influences of their surroundings. One can easily see how this attitude might have encouraged the appeal to conceptual models borrowed from the macroscopic domain, such as the four engineering metaphors I have considered in this chapter. Indeed, Schrödinger himself, as I have just discussed, acknowledged the deep resemblance between organisms and machines with respect to the kind of order they exhibit, as well as to the negligible impact of the physical environment on their operation.

Some fairly compelling evidence for this hypothesis can be found by considering the case of Monod, one of the main intellectual architects of the molecular biology revolution. The reason is that Monod appears to have changed his mind about the nature of biological order partially as a consequence of reading *What is Life?* which he regarded as a work of genius (see Loison 2015). Though initially committed earlier in his career to a statistical and non-deterministic understanding of biological regularities (consistent with the aforementioned order-from-disorder principle), Monod later came to regard the order of the cell as a product of the static, clockwork-like precision of its macromolecular components. "The whole trend of modern molecular biology," Monod declared in a 1958 lecture, "makes it every day clearer that structural stability and rigidity rather than dynamicity are the most essential and characteristic properties of the typical cellular macromolecules" (Monod, quoted in Loison 2015, 395). Monod also commented in his notes for that same lecture, where he explicitly mentioned Schrödinger in a parenthetical remark, that even when examining large macromolecules (e.g., ribosomes) and complex subcellular processes (e.g., protein synthesis), one can confidently disregard the disruptive effects of stochasticity due to the imposing stability and rigidity of the participating molecules:

> The protein-synthesizing process appears to work with very high precision, and the concept of molecular micro-heterogeneity due to errors or fluctuations in this process appears unwarranted. Putting it otherwise: even in the formation of such a very large and complex molecule [i.e., a protein], the synthesizing system [i.e., the ribosome] appears to work mechanically, like a clock or a precision machine tool, rather than statistically (Schrödinger).
>
> (Monod, quoted in Loison 2015, 396)[10]

By the time he wrote his renowned treatise on molecular biology, *Chance and Necessity*, Monod had become even more forceful in his dismissal of stochastic environmental effects, noting that "a living being's structure [. . .] owes almost nothing to the action of outside forces, but everything, from its overall shape down to its tiniest detail, to [. . .] interactions within the object itself" (Monod 1972, 10). Note that this is precisely the view that I have repeatedly challenged over the course of this chapter.

In any case, more historical research is needed to corroborate this interpretation. Still, I cannot resist making the provocative observation that if the proposed hypothesis is correct, we shall be forced to draw the utterly paradoxical conclusion that one of the greatest physicists of the 20th century was responsible for making several generations of molecular biologists forget about the importance of *physical forces* on the phenomena they study.

## Conclusions

Despite its importance in the historical development of biological thought, the machine conception of the organism is deeply problematic from an ontological point of view. In this chapter I have drawn on the insights that Haldane offered in a classic essay from the 1920s to propose a new philosophical argument against this conception that is particularly relevant for current work in cell and molecular biology, and which I have called the Argument from Scale. This states that, owing to their minuscule size, cells and their macromolecular components are subject to drastically different physical conditions compared with macroscopic objects like machines, and that using machine metaphors to explain microscopic phenomena is consequently more likely to obscure and deceive than it is to elucidate and enlighten. I have illustrated this argument by analyzing four central conceptual models in molecular biology that were originally imported from electronic and mechanical engineering – genetic program, cellular circuitry, molecular machine, and molecular motor – and by showing that their explanatory deficiencies ultimately derive from their neglect of the impact of scale. Once scale is seriously taken into account, it becomes hard to defend the theoretical adequacy of these models (which, of course, is not to say that they cannot sometimes serve useful heuristic purposes as convenient, experimentally tractable idealizations).

Although there will be many that will continue to believe that "[t]he engineering sciences, particularly electronic and control engineering, are likely to have an ever increasing and pervasive impact on molecular biology" (Sauro and Kholodenko 2004, 37), the fact is that the physical dimensions of the cell and the milieu it finds itself in impose fundamental constraints on what is possible and what is not, both structurally and functionally. The rigidity, stability, and deterministic precision that we typically associate with the machines of our macroscopic world simply cannot exist in the messy, turbulent, and chaotic world that cells and molecules inhabit. "The clockwork mechanism of the cell," if that is what we insist on calling it, is "built not of precisely engineered solid cogs, but of vague and uncertain particles whose generation, diffusion, and reaction can not provide any precision"

(Hallett 1997, 105). Perhaps the most serious obstacle in coming to terms with molecular and cellular phenomena is the lack of a good analogy from our daily experience. It is for this reason that we should learn to trust what physics tells us about the molecular realm, despite being strange and counterintuitive, over the more familiar and comforting picture that traditional appeals to engineering have tended to suggest.

## Acknowledgments

I wish to warmly thank Sune Holm for inviting me to participate in the extremely stimulating workshop on the topic of "living machines" held at the Carlsberg Academy in Copenhagen, where this paper was first presented. In addition, I received valuable feedback on other presentations of this paper from audiences at the *Mechanism and Autonomy in Biology* workshop in Kassel, the annual UK integrated history and philosophy of science conference in Nottingham, the *Size, Development, and Evolution* symposium in Madrid, the *Theoretical Underpinnings of Molecular Biology* summer school in Rijeka, and the second *PhilinBioMed* meeting in Bordeaux. I also thank Adam Wilkins, Brett Calcott, and my colleagues at the Konrad Lorenz Institute for Evolution and Cognition Research for very helpful comments on earlier drafts of this chapter.

## Notes

1 For a more technical and detailed examination of the numerous theoretical problems with the machine conception of the cell that complements and extends the analysis I provide in this chapter, see Nicholson 2019.

2 To a terrestrial one, at any rate. The threat that gravity poses to a large aquatic animal (such as a whale) is greatly reduced, as it can use its buoyancy to counteract gravitational effects.

3 Nor was he the last. A highly accessible contemporary account of the impact of size in biology can be found in Bonner 2006.

4 In a fascinating historical examination of the genesis of the genetic program metaphor, Peluffo (2015) explores the potential intellectual connections between Jacob and Monod on the one hand and Mayr on the other prior to their respective 1961 publications.

5 The genetic program is not, of course, the only metaphor biologists have used to characterize gene expression or embryonic development. For a discussion of alternative, non-machine-based conceptualizations of these processes, see Nicholson 2014.

6 Even enzymes, which have traditionally been regarded as extremely specific catalysts, exhibit varying degrees of catalytic promiscuity, as well as the ability to perform a wide range of non-catalytic functions, including cell motility, membrane trafficking, chaperoning, activation and inhibition of metabolic pathways, and chromatin organization (Babtie et al. 2010; Khersonsky and Tawfik 2010).

7 In a recent paper, Militello and Moreno (2018) have defended the legitimacy of the term "molecular machine" in the characterization of macromolecular assemblies – even after recognizing their patently nonmechanical features – by proposing to define a machine as "a meta-stable structure consisting of interdependent parts which constrain a flow of energy and matter in order to do work and perform a systemic function" (ibid.: 35) and showing that this definition can accommodate what we know

about these subcellular entities. While I agree that such a broad definition is capacious enough to encompass many (though perhaps not all) macromolecular assemblies in the cell, my own preference, following Mayer et al. (2009) and others, is to embrace an alternative, explicitly nonmechanical conceptualization of them as *pleomorphic ensembles* (for details, see Nicholson 2019).

8 Indeed, it is highly unlikely that a given ribosome ever repeats its exact same movements as it elongates a polypeptide.

9 Schrödinger illustrates this with the example of the "Habsburg lip," a genetic trait afflicting the Habsburg dynasty that persisted for hundreds of years despite having a molecular basis and consequently being permanently subject to the turbulence of thermal agitation.

10 Compare Monod's characterization of the operation of the ribosome with the radically opposing one offered half a century later by Moore (2012), which I quoted earlier in this chapter. The contrast between the two is extraordinary.

## References

Ait-Haddou, R., & Herzog, W. (2003). Brownian ratchet models of molecular motors. *Cell Biochemistry and Biophysics*, 38, 191–212.

Alberts, B. (1998). The cell as a collection of protein machines: Preparing the next generation of molecular biologists. *Cell*, 92, 291–294.

Alberts, B., Johnson, A., Lewis, J., Raff, M., Roberts, K., & Walter, P. (2008). *Molecular Biology of the Cell*, 5th ed. New York: Taylor and Francis.

Altschuler, S. J., & Wu, L. F. (2010). Cellular heterogeneity: Do differences make a difference? *Cell*, 141, 559–563.

Asbury, C. L. (2005). Kinesin: World's tiniest biped. *Current Opinion in Cell Biology*, 17, 89–97.

Astumian, R. D. (2001). Making molecules into motors. *Scientific American*, 285, 56–64.

Astumian, R. D. (2007). Design principles for Brownian molecular machines: How to swim in molasses and walk in a hurricane. *Physical Chemistry Chemical Physics*, 9, 5067–5083.

Atlan, H., & Koppel, M. (1990). The cellular computer DNA: Program or data. *Bulletin of Mathematical Biology*, 52, 335–348.

Babtie, A., Tokuriki, N., & Hollfelder, F. (2010). What makes an enzyme promiscuous? *Current Opinion in Chemical Biology*, 14, 200–207.

Balázsi, G., van Oudenaarden, A., & Collins, J. J. (2011). Cellular decision making and biological noise: From microbes to mammals. *Cell*, 144, 910–925.

Bier, M. (2003). Processive motor protein as an overdamped Brownian Stepper. *Physical Review Letters*, 91, 148104-1-148104-4.

Blinov, M. L., & Moraru, I. I. (2012). Logic modeling and the ridiculome under the rug. *BMC Biology*, 10(92), 1–8.

Block, S. M. (1997). Real engines of creation. *Nature*, 386, 217–219.

Bonner, J. T. (2006). *Why Size Matters: From Bacteria to Blue Whales*. Princeton: Princeton University Press.

Bray, D. (2009). *Wetware: A Computer in Every Cell*. New Haven: Yale University Press.

Browne, W. R., & Feringa, B. L. (2006). Making molecular machines work. *Nature Nanotechnology*, 1, 25–35.

Canguilhem, G. (2008). *Knowledge of Life*. New York: Fordham University Press.

Copley, S. D. (2003). Enzymes with extra talents: Moonlighting functions and catalytic promiscuity. *Current Opinion in Chemical Biology*, 7, 265–272.

Danchin, A. (2009). Bacteria as computers making computers. *FEMS Microbiology Reviews*, 33, 3–26.

Davidson, E. H. (2001). *Genomic Regulatory Systems: Development and Evolution*. San Diego: Academic Press.

Davidson, E. H. (2009). Network design principles from the sea urchin embryo. *Current Opinion in Genetics and Development*, 19, 535–540.

Davidson, E. H., & Levine, M. (2005). Gene regulatory networks. *Proceedings of the National Academy of Sciences*, 102, 4935.

Davidson, E. H., McClay, D. R., & Hood, L. (2003). Regulatory gene networks and the properties of the developmental process. *Proceedings of the National Academy of Sciences*, 100, 1475–1480.

Dueber, J. E., Yeh, B. J., Bhattacharyya, R. P., & Lim, W. A. (2004). Rewiring cell signaling: The logic and plasticity of eukaryotic protein circuitry. *Current Opinion in Structural Biology*, 14, 690–699.

Dumont, J. E., Pécasse, F., & Maenhaut, C. (2001). Crosstalk and specificity in signalling: Are we crosstalking ourselves into general confusion? *Cellular Signalling*, 13, 457–463.

Eldar, A., & Elowitz, M. B. (2010). Functional roles for noise in genetic circuits. *Nature*, 467, 167–173.

Elowitz, M. B., Levine, A. J., Siggia, E. D., & Swain, P. S. (2002). Stochastic gene expression in a single cell. *Science*, 297, 1183–1186.

Frank, J. (2000). The ribosome: A macromolecular machine par excellence. *Chemistry and Biology*, 7, R133–R141.

Frank, J. (ed.). (2011). *Molecular Machines in Biology*. Cambridge: Cambridge University Press.

Fuxreiter, M., & Tompa, P. (eds.). (2012). *Fuzziness: Structural Disorder in Protein Complexes*. New York: Springer.

Galileo (1914 [1638]). *Dialogues Concerning Two New Sciences*. New York: Macmillan.

Garrett, J. (1999). Mechanics of the ribosome. *Nature*, 400, 811–812.

Grmek, M. D. (1972). A survey of the mechanical interpretations of life from the Greek atomists to the followers of Descartes. In A. D. Breck & W. Yourgrau (eds.), *Biology, History, and Natural Philosophy*, 181–195. Boston, MA: Springer.

Gierasch, L. M., & Gershenson, A. (2009). Post-reductionist protein science, or putting humpty dumpty back together again. *Nature Chemical Biology*, 5, 774–777.

Haldane, J. B. S. (1928). *Possible Worlds and Other Papers*. New York: Harper & Brothers.

Hallett, M. B. (1997). Is "Life" based on clockwork biology or quantum uncertainty? *Perspectives in Biology and Medicine*, 41, 101–107.

Hanahan, D., & Weinberg, R. A. (2000). The hallmarks of cancer. *Cell*, 100, 57–70.

Howard, J. (2001). *Mechanics of Motor Proteins and the Cytoskeleton*. Sunderland: Sinauer Associates.

Howard, J. (2006). Protein power strokes. *Current Biology*, 16, R517–R519.

Jacob, F. (1973). *The Logic of Life*. New York: Pantheon.

Jacob, F., & Monod, J. (1961). Genetic regulatory mechanisms in the synthesis of proteins. *Journal of Molecular Biology*, 3, 318–356.

Jeffery, C. J. (2003). Moonlighting proteins: Old proteins learning new tricks. *Trends in Genetics*, 19, 415–417.

Keller, E. F. (1995). *Refiguring Life: Metaphors of Twentieth-Century Biology*. New York: Columbia University Press.

Keller, E. F. (2000). Decoding the genetic program: Or, some circular logic in the logic of circularity. In P. J. Beurton, R. Falk, & H.-J. Rheinberger (eds.), *The Concept of the Gene in Development and Evolution*, 159–177. Cambridge: Cambridge University Press.

Khersonsky, O., & Tawfik, D. S. (2010). Enzyme promiscuity: A mechanistic and evolutionary perspective. *Annual Review of Biochemistry*, 79, 471–505.

Knight, H., & Knight, M. R. (2001). Abiotic stress signalling pathways: Specificity and cross-talk. *Trends in Plant Science*, 6, 262–267.

Kupiec, J-J. (2010). On the lack of specificity of proteins and its consequences for a theory of biological organization. *Progress in Biophysics and Molecular Biology*, 102, 45–52.

Kurakin, A. (2009). Scale-free flow of life: On the biology, economics, and physics of the cell. *Theoretical Biology and Medical Modelling*, 6(6), 1–28.

Lewontin, R. C. (2000). *The Triple Helix: Gene, Organism, and Environment*. Cambridge, MA: Harvard University Press.

Linke, H., Downton, M. T., & Zuckermann, M. J. (2005). Performance characteristics of Brownian motors. *Chaos*, 15, 026111.1–026111.11.

Loison, L. (2015). Why did Jacques Monod make the choice of mechanistic determinism? *Comptes Rendus Biologies*, 338, 391–397.

Longo, G., & Tendero, P-E. (2007). The differential method and the causal incompleteness of programming theory in molecular biology. *Foundations of Science*, 12, 337–366.

Mayer, B. J., Blinov, M. L., & Loew, L. M. (2009). Molecular machines or pleiomorphic ensembles: Signalling complexes revisited. *Journal of Biology*, 8, 1–8.

Mayr, E. (1961). Cause and effect in biology. *Science*, 134, 1501–1506.

Mayr, E. (1982). *The Growth of Biological Thought*. Cambridge, MA: Harvard University Press.

Militello, G., & Moreno, A. (2018). Structural and organisational conditions for being a machine. *Biology & Philosophy*, 33, 35.

Misteli, T. (2001). Protein dynamics: Implications for nuclear architecture and gene expression. *Science*, 291, 843–847.

Monod, J. (1972). *Chance and Necessity: An Essay on the Natural Philosophy of Molecular Biology*. New York: Vintage.

Moore, P. B. (2012). How should we think about the ribosome? *Annual Review of Biophysics*, 41, 1–19.

Moss, L. (1992). A kernel of truth? On the reality of the genetic program. *PSA: Proceedings of the Biennial Meeting of the Philosophy of Science Association*, 1, 335–348.

Nicholson, D. J. (2013). Organisms ≠ Machines. *Studies in History and Philosophy of Biological and Biomedical Sciences*, 44, 669–678.

Nicholson, D. J. (2014). The machine conception of the organism in development and evolution: A critical analysis. *Studies in History and Philosophy of Biological and Biomedical Sciences*, 48, 162–174.

Nicholson, D. J. (2018). Reconceptualizing the organism: From complex machine to flowing stream. In D. J. Nicholson & J. Dupré (eds.), *Everything Flows: Towards a Processual Philosophy of Biology*, 139–166. Oxford: Oxford University Press.

Nicholson, D. J. (2019). Is the cell *really* a machine? *Journal of Theoretical Biology*, 477, 108–126.

Nijhout, H. F. (1990). Metaphors and the role of genes in development. *BioEssays*, 12, 441–446.

Nobeli, I., Favia, A. D., & Thornton, J. M. (2009). Protein promiscuity and its implications for biotechnology. *Nature Biotechnology*, 2, 157–167.

Nogales, E. (2016). The development of cryo-EM into a mainstream structural biology technique. *Nature Methods*, 13, 24–27.

Nogales, E., & Grigorieff, N. (2001). Molecular machines: Putting the pieces together. *Journal of Cell Biology*, 152, F1–F10.

Oster, G., & Wang, H. (2003). How protein motors convert chemical energy into mechanical work. In M. Schliwa (ed.), *Molecular Motors*, 207–227. Weinheim: Wiley.

Oyama, S. (2000). *The Ontogeny of Information*, 2nd ed. Durham, NC: Duke University Press.

Peluffo, A. E. (2015). The "genetic program": Behind the genesis of an influential metaphor. *Genetics*, 200, 685–696.

Piccolino, M. (2000). Biological machines: From mills to molecules. *Nature Reviews*, 1, 149–153.

Rao, C. V., Wolf, D. M., & Arkin, A. P. (2002). Control, exploitation and tolerance of intracellular noise. *Nature*, 420, 231–237.

Raser, J. M., & O'Shea, E. K. (2005). Noise in gene expression: Origins, consequences, and control. *Science*, 309, 2010–2013.

Reynolds, A. (2018). *The Third Lens: Metaphor and the Creation of Modern Cell Biology*. Chicago: Chicago University Press.

Rosenberg, A. (1997). Reductionism redux: Computing the embryo. *Biology and Philosophy*, 12, 445–470.

Rueda, M., Ferrer-Costa, C., Meyer, T., Pérez, A., Camps, J., Hospital, A., Gelpi, J. L., & Orozco, M. (2007). A consensus view of protein dynamics. *Proceedings of the National Academy of Sciences*, 104, 796–801.

Sauro, H. M., & Kholodenko, B. N. (2004). Quantitative analysis of signaling networks. *Progress in Biophysics and Molecular Biology*, 86, 5–43.

Schrödinger, E. (1944). *What Is Life? The Physical Aspect of the Living Cell*. Cambridge: Cambridge University Press.

Thompson, D. W. (1942). *On Growth and Form*, 2nd ed. Cambridge: Cambridge University Press.

Tyska, M. J., & Warshaw, D. M. (2002). The myosin power stroke. *Cell Motility and the Cytoskeleton*, 51, 1–15.

Uversky, V. N. (2013). Unusual biophysics of intrinsically disordered proteins. *Biochimica et Biophysica Acta*, 1834, 932–951.

Vale, R. D., & Milligan, R. A. (2000). The way things move: Looking under the hood of molecular motor proteins. *Science*, 288, 88–95.

Vartanian, A. (1973). Man-machine from the Greeks to the computer. In P. Wiener (ed.), *Dictionary of the History of Ideas*, Vol. 3, 131–146. New York: Scribner.

Wang, B., & Buck, M. (2012). Customizing cell signaling using engineered genetic logic circuits. *Trends in Microbiology*, 20, 376–384.

Webster, G., & Goodwin, B. C. (1982). The origin of species: A structuralist approach. *Journal of Social and Biological Structures*, 5, 15–47.

Wolpert, L. (1994). Do we understand development? *Science*, 266, 571–572.

Wright, P. E., & Dyson, H. J. (2015). Intrinsically disordered proteins in cellular signalling and regulation. *Nature Reviews Molecular Cell Biology*, 16, 18–29.

Xie, X. S., Choi, P. J., Li, G-W., Lee, N. K., & Lia, G. (2008). Single-molecule approach to molecular biology in living bacterial cells. *Annual Review of Biophysics*, 37, 417–444.

Yang, H., Luo, G., Karnchanaphanurach, P., Louie, T-M., Rech, I., Cova, S., Xun, L., & Xie, X. S. (2003). Protein conformational dynamics probed by single-molecule electron transfer. *Science*, 302, 262–266.

Zhou, Y., Vitkup, D., & Karplus, M. (1999). Native proteins are surface-molten solids: Application of the Lindemann criterion for the solid versus liquid state. *Journal of Molecular Biology*, 285, 1371–1375.

# 3 A roomful of robovacs

## How to think about genetic programs

*Brett Calcott*

## Introduction

> *The genomic sequence encodes the developmental program which determines the progression from fertilized egg to organized body plan.*
>
> *(Peter and Davidson 2013, 75)*

Peter and Davidson's statement is not unusual; biologists frequently equate or compare genes to computer programs. Their aim is often to identify a special or distinctive role for genes in development. Thus, genes contain the instructions for building an organism in contrast to, say, mere environmental inputs. Despite the frequency of these statements, many philosophers and biologists argue that such comparisons are wrong-headed and ultimately misleading rather than informative (Nicholson 2014; Keller 2001; Pigliucci 2010; Boudry and Pigliucci 2013; Griffiths 2001; Planer 2014; Nijhout 1990; Moczek et al. 2015; Jaeger et al. 2015). The complaints are many, and I won't rehearse them all here. Instead, I want to examine an aspect of the comparison between genes and programs that has had scant attention from philosophers.

The issue is this: if we wish to advocate or dismiss the idea that genes are like a program, then we need a clear idea about what a program itself is like. Yet while the debates about genetic programs draw on the latest insights from molecular and developmental biology, the corresponding talk about programs provides little detail about what programs are and how they function; it is simply assumed that this is common wisdom. But the discussion typically centers on a very narrow interpretation of how programs operate. For example, the debates often assume that a program is a list of instructions telling the computer what to do. This is true of some programming languages (such as C, Java, or Python), but in other languages (such as SQL or Haskell), you describe the results required rather than dictate the instructions and the order they should be done. Many claims in the debates rely on one understanding of a program, rather than anything about programs in general (see Backus 1978 for a critique of these assumptions).

For many critics, this narrow interpretation of programs is irrelevant. Their goal has been to evaluate the use and misuse of the genetic programming metaphor in recent biological history (Keller 2003, for example). Their critiques often hit the mark because advocates of genetic programs deploy concepts that are either similarly narrow or vague enough to be interpreted this way.

But evaluating how scientists have deployed the notion of program is not our only option here. An alternative is to look beyond this narrow interpretation and see if it is possible to build a better analogy. Why bother? One role that engineering analogies play in science is to draw on something well understood to illuminate, clarify, or rethink some puzzling natural phenomenon. If we discard the genetic program analogy without fully exploring what programs are like, we risk losing valuable insights or simply reinventing them under another name.

In the rest of this chapter, I sketch a way to think about programs that avoids many of the criticisms advanced against genetic programs. This suggests there is still some worth in thinking of genes as programs, as long as we're clear on what programs we have in mind.

## Avoiding deep thought

In Douglas Adams's tale *The Hitchhiker's Guide to the Galaxy*, intelligent mice construct a computer named Deep Thought to compute the answer to the meaning of life, the universe, and everything. The mice then wait 7.5 million years for it to deliver its famous answer. That is an absurdly long time to wait, but it highlights something often attributed to computers and, by extension, programs. The computational task demanded of Deep Thought has what we might call a "ballistic trajectory." We give the computer (or program) some input, set it running, and then stand back and wait for it to produce an answer. The British "bombe" device depicted in the film about Alan Turing (*The Imitation Game*, 2014) demonstrates this same ballistic trajectory, whirring and clicking to crack the German enigma code (just in time, whew!).

Many programs have this ballistic property. When biologists use phylogenetic software to reconstruct the tree-like relationships between organisms using their DNA, they feed it the raw sequence information and then may wait weeks for these programs to produce results. A program need not be complex or long-running to be ballistic, however. Consider the following exercises that might be found in a beginners course on programming:

- Exercise 1: write a program to compute the first N prime numbers.
- Exercise 2: write a program that calculates someone's age in days from a given birth date (look out for leap years!).

In each exercise, we're asked to write a program that begins by consuming some input, performs some calculation, and ends by supplying some output. A program like this, which starts with a set of initial conditions and then churns through an ordered series of steps to produce a final output, is sometimes known as a "script" or a "batch program."

Most programs we use today are not batch programs. When I start my word processor, I don't sit around waiting for it to finish its job. The opposite is often true: my word processor is sitting around waiting for me. A word processor, like most programs we encounter these days, is an "interactive program": it takes in various bits of input from the keyboard and mouse, responding with changes to what appears on the screen and the odd beeping noise. Each interaction is brief and evokes a rapid response: a keystroke adds a letter; a menu selection formats the text. A long series of such interactions produce a story or a business report or a scientific paper.

This interplay between the program and user describes a feedback loop. I tap a few words, the word processor highlights a spelling mistake, and I go back and correct it. I read over a few sentences, decide on a better formulation, and then go back and clarify the text. The dynamic coupling between the user and program is a mark of these interactive programs.

The dynamic coupling of the interaction is essential, for interactive programs are not merely programs that require input at times other than when they are started. Consider an installation program from the days of CDs and floppy disks. These programs stopped intermittently with the tedious but necessary request: "please insert next disk." Such a program is interactive in one sense – it asks for something from a user at various stages during its operation. But notice that, in this case, the *program decides* when and what type of interaction takes place. With a word processor, the user is in the driver's seat, and the program awaits commands in the form of mouse clicks and key presses. Something outside the program (us, in this case) decides the order things happen rather than the other way around.

The distinction between batch program and interactive program is not a hard line; a single program might contain aspects of both. For example, we might view commands executed by a word processor, such as "Print" or "Format Bibliography," as tiny batch subprograms in themselves; we start them up and wait for them to finish. But the difference between batch programs and interactive programs is relevant when we make analogies with genes, for it colors our thinking about what programs are, how they operate, and what they explain.

## Which program are you thinking of?

Critics of the genomic program often appeal to features of programs that apply only to batch programs. For example, Evelyn Fox Keller thinks of a program as something that starts with some data and finishes by producing some output – following the ballistic trajectory I outlined earlier. Because of this, her only way of envisioning complex gene interaction with feedback is to invoke multiple programs:

> [. . .] what counts as "data" for one "program" is often the output of a second "program," and the output of the first is "data" for yet another "program," or even for the very "program" that provided its own initial "data."
>
> (Keller 2001)

Yet the notion of a feedback loop in interactive programs renders this unnecessary. A single interactive program can take in input at different times, and that input may include output that it has produced at previous times.

Fred Nijhout, similarly, appears to have batch programs in mind in his critique of the genetic program metaphor. He outlines two conditions for gene expression to be a program (Nijhout 1990, 442):

1    "The gene or its product must be necessary and sufficient for the occurrence of the process, and not be itself provoked by the process itself."
2    "A program must somehow contain information about the temporal sequence of events."

The first claim sounds reasonable for something like a batch program: we start up a program to construct a phylogenetic tree, give it some input, and then let it run. After setting it going, the results are determined by the program alone. Now consider an interactive program: a half-written document, displayed on the screen, provokes the user to change it, perhaps by pointing out a spelling mistake. So the feedback loop between user and program guarantees it will fail Nijhout's first condition.

Nijhout's second condition fares the same. A batch program comprises a set of modular components *and* some order in which to execute these components. The order is important, for the operations later in the program often depend on the completion of earlier operations. An interactive program also has a set of modular components. In a word processor, the various menu options expose some of this modular functionality (such as saving, formatting, numbering, or printing). But the order in which we put together these operations – the "temporal sequence of events" – is not part of the program. The user of the program chooses how to put them together.

Both of Nijhout's conditions draw on a more basic assumption – that a program contains some intrinsic ordering over how the instructions in the program are executed. Nijhout is not alone in emphasizing the importance of this intrinsic ordering. Another critic of the genetic program, Ronald Planer, finds it problematic that "there is no order in which these instructions can be properly said to be retrieved and executed by the cell during development," so that there is no "beginning, middle, or end to this 'program'" (Planer 2014).

Consider Planer's claim in light of batch programs and interactive programs. A batch program has a clear beginning, middle, and end. Many batch programs even show how far through the various tasks they are: "Processing 50% complete . . ." But how should we respond if asked about the beginning, middle, and end of a word processing program? Given the technical know-how, we might find the first instructions executed by a word processor when it starts and maybe even the last instruction it executes when we quit the program. But where is the middle of this program? The program comprises a set of small actions – typing, formatting, editing – chained together in a variety of ways by the user of the program. There is no way to point to some specific piece of code and say, "This is the middle," like we could with, say, the program that reconstructs phylogenies. This is because an interactive program does not dictate the order in which it must execute its operations. Instead, this order derives from some external input.

Notice that a word processor is useful *because* it does not dictate this order. Without the ability to control the order in which it executes the various components, the flexibility of the word processor would be lost. This exposes assumptions about what programs are good for. If we focus on batch programs, we might think the utility of a program lies in its capacity to encode a complex series of dependent operations. With a word processor, however, rather than a set of dependent steps, we have a set of loosely coupled operations that transform a document. The utility of the program lies in the ease and flexibility with which some external process can recombine these operations. An interactive program resembles a well-designed toolkit of related operations that are assembled at run time rather than a set of ordered instructions that execute a well-designed plan.

This shift in focus turns many common ideas related to programs, and especially *genetic* programs, on their head (see Nicholson 2014, for example). A batch program emphasizes determinism; an interactive program, flexibility and open-endedness. A batch program churns away by itself; an interactive program depends on inputs from some external source. A batch program contains the rules (instructions, procedures, algorithm) for producing its output; an interactive program contains no rules for producing a particular output but a toolkit for generating many outputs. Finally, a batch program often incites agentive thinking: "the phylogenetic program *generated* a tree of related various organisms," while an interactive program does not: Microsoft Word no more *wrote* my thesis than a typewriter *wrote* Hemingway's *The Old Man and the Sea*.

## No spookiness required

I introduced an interactive program using a word processor – something I assume readers are familiar with. But a word processor interacts with *us* – an intelligent external agent. Am I suggesting there is an external intelligent agent interacting with (or directing!) the genetic program? That would be spooky. So let me describe an interactive program without positing any intelligent external agent.

Instead of a word processor, consider a program that controls a mobile robot, such as a robot vacuum cleaner or "robovac." If you haven't seen one, they are self-contained vacuum cleaners the size and shape of a Frisbee on wheels. They zip around the floor, moving underneath furniture and doing a surprisingly good job of cleaning your house unattended.

A robovac has several sensors that convey information about the local environment – whether it is touching something, whether something is in front, and so on. This information feeds into the program controlling the robovac, and the program responds by maneuvering the robot, such as changing direction or slowing down. By maneuvering, the robot changes its environment, and now its sensors are exposed to a different environment. This results in further maneuvering. And so it goes until (in my experience) the battery runs out or the robovac ingests a stray sock. Here we have the same coupled, fluid feedback loop between a program and some external set of affairs. But instead of an intelligent agent directing the robot, it is the changing environment that serves as the other half of the feedback loop.

The robovac example removes the spookiness, but have we lost the contrast with the batch program I outlined earlier? Unlike the word processor's open-ended output, a robovac's actions are directed to one outcome: cleaning the house. Let's shift our attention from this goal and consider, instead, how the robovac achieves this task. Take the path the robovac robot traced around my house on some particular day. This path was not programmed into the robot – I did not upload a house plan into my robovac and have it precalculate some optimal cleaning path. Rather, the particular path taken arose from the continuous interaction between the layout of my house and the combined set of responses of the program controlling the robot's actions.

This "emergent" behavior is often contrasted with how programs work.[1] Nijhout, for example, differentiates development from a program by describing it thus: "Development is a series of elaborate temporal and spatial interactions that are context dependent. The sequence of gene activation we see in development is an emergent property of this interaction" (Nijhout 1990). Yet we can describe the sequence of maneuvers executed by our robovac the same way. The particular path the robovac took is an emergent property of its interaction with the environment. Like the word processor, the ordering of the set of responses it made was not intrinsic but induced by something external to the robot. For the robovac, this external feature was the local environment. This local environment, in turn, resulted from previous decisions made by the robot.

Switching the kind of program we have in mind changes what features we attribute to a program. Yet it is batch programs that critics have focused on, while interactive programs look like a better choice if we wish to build an analogy with genes. In the next section, I show how extending this idea can add some further clarity to constructing a useful analogy.

## A roomful of robovacs

Word processors and robovacs are interactive programs, but what they interact with differs. In the first case, the program interacts with a goal-directed intelligent agent (such as me, on good days at least). But a robovac largely interacts with a *static* environment, a variety of furniture distributed across rooms of various shapes. Dynamic bits of the environment, such as pets and small children, typically present a challenge to it. The variety of input it receives is generated entirely by its own motion in that environment.

Now consider placing several robovacs in the same room. We no longer have a static environment, as each robovac is responding, in part, to other robovacs. To my knowledge, no one has designed robovacs that work in groups, so I doubt this exercise would produce faster or more efficient cleaning (over and above there being just *more* robovacs). But this kind of collective robotic behavior is being actively explored. There are flocks of coordinated aerial robots that can fly in formation, for example (Vasarhelyi et al. 2014). These flocks lack any central controlling system; each robot is independent, responding to local physical parameters, global positioning information, and messages received from other robots. Their ability to fly in formation – a group-level ability – results from the continuous interaction between individual robots.

Would we say the program *created* a particular flying formation? That seems odd but for a different reason than saying the word processor *wrote* my thesis. For the question here is about a group-level capacity: a particular flying formation. A single robot (with a single program) does not create a formation. If we are to attribute this goal to the program, then we need to mention it arose from several copies of that program interacting with one another. We can think of it like this. For the robots to assemble into a formation, each must maneuver itself while continually adjusting to the other robots doing likewise. As with our initial robovac, these paths are not explicitly coded into the robots. Rather, they emerge from each robot's coupled interaction with other robots navigating their own path. But the maneuvering of individual robots is not equivalent to the assembly of a particular formation. *That* emerges from the collective behavior of all the robots.

So neither the behavior of individual aerial robots nor their group flying formation is encoded in the program. To fully understand how the formation occurs, we need to understand how the program is embedded within the physical context and how the communication and interaction between the multiple versions of it take place. Yet we can still make sense of a program playing a distinctive role in *controlling* a robot. The interactive nature of program is, in part, what enables the robot to respond with appropriate behavior given particular local conditions.

Let me summarize the kind of program we are now dealing with. I'll call it a Collectively-Identical Interactive Program. Abstracting away from the details of our roomful of robovacs (or skyful of aerial robots), we have:

1   A single interactive program (a related but loosely coupled set of operations, which are invoked by, and respond to, incoming input),
2   where there are multiple identical copies of this same program running concurrently,
3   and each copy of the interactive program is coupled to other copies of itself, where the outputs from one can affect the future inputs of others.
4   Furthermore, the program has been designed to interact with other copies of itself,
5   yet the design goal itself is a group-level capacity rather than the individual behavior of a single copy of the program.

This kind of program naturally fits with the goal of controlling swarms of robots or agent-based simulations. But this is not the only place we find this confluence of properties. One recent example concerns analyzing the social networks that arise in online interactions – a task essential for many big online companies. These social networks comprise "vertices," representing people, and "edges," representing social relationships (capturing friendship, for example). Understanding the structure of these networks can reveal an enormous amount about how people interact (too much perhaps). It turns out, however, that traditional top-down algorithms for analyzing networks cannot deal with networks of this size. One solution to this problem, proposed by employees at Google, is to treat the graph as though each vertex (or node) is running the same program and receiving input and sending output along its edges (Malewicz et al. 2010). Because each program requires only limited local information, we can distribute these programs across many computers and run in parallel. Similar to the way the iterative feedback between programs in interacting aerial robots converges to a particular formation, we can use iterative feedback between many interacting vertex-based programs to identify high-level structure in graphs, such as the complex communities of friendships in a social graph. There are no robots here, but we still have a program that interacts with itself to produce a collective feature at a higher level.

## Building a better analogy

Given the aim of this chapter, it should not be surprising that I think the kind of program I have just outlined has something in common with genes. The analogy goes like this. In a developing multicellular organism, each cell contains a copy of the same DNA.[2] This DNA encodes a program, and each cell is running a copy of the same program. These are interactive programs, constantly responding to their local environment. In a developing multicellular organism, this environment consists largely of other cells. Like the examples, each copy of the program interacts with other copies in the surrounding cells. On this view, a genomic program is an interactive program that has evolved to interact with copies of itself to generate a higher level of organization, such as prominent features of a multicellular body plan.

   This claim might sound familiar and open to the same familiar critiques. But it is essential to see these claims in the light of the preceding discussion about programs. First, although we might say (one copy of) a program is guiding the cell's behavior, this behavior is not encoded as a series of steps in the genome but arises dynamically from the interaction between the program and its changing developmental environment. Second, not only is this cellular behavior not encoded in the genome, but the individual cellular behavior itself is also not equivalent to the higher level of organization that the collective individual-level behavior produces.

   This revised view of a genomic program allows us to pull apart some features that are commonly run together. For example, it makes sense to say that this program controls features of development because changing the program will change how the collective interaction of the programs unfolds.[3] But it does not follow that the program itself is sufficient to "compute the embryo." Nor does it mean that this higher-level organization is "in" the DNA or that it can be read off the DNA like the structure of a building can be read off a blueprint. This version of a genetic program is also compatible with views that emphasize the key role of physical self-organization in development (such as Newman and Bhat 2009). As we saw with the aerial robots, the programs by themselves did not explain how the formations were constructed; other information about the interactions among the group was required.

## Summary

My goal in this chapter was to present a fresh approach to genetic programs. I focused on two key ideas. First, we need to move beyond scripts and batch programs and instead look at how interactive programs work. Second, we need to look at programs that are specifically designed to interact with one another to produce some collective behavior. The first of these is commonplace – almost every program we now use is interactive. The second shift is more specialized, but I identified two areas of active research. More can be said, however. I've focused on describing the behavioral contrasts between batch and interactive programs. But the internal architecture of these programs differs too. I touched on this earlier when I referred to an interactive program as more like a combinatorial toolkit whose pieces are put together at run time. Interestingly, this same architecture enables interactive programs to be extended and modified more easily than batch programs. That looks relevant once we shift our attention to the evolution of development and the subject of evolvability (Calcott 2014).

The notion of a genetic program still has something to offer us. It may not be a complete picture, nor should it be the only source of ideas. It does, however, have additional virtues in contrast to frameworks offered as replacements, such as developmental systems theory (Oyama et al. 2003). For when you borrow from engineering, the ideas you draw on have been applied and are known to work, and their limitations are understood.

## Notes

1 By "emergent," here (and elsewhere), I mean the weak emergence exhibited by many agent-based systems and cellular automata rather than anything metaphysically hard to understand (Bedau 1997).
2 To a rough approximation. There are always exceptions in biology, such as cells without DNA, somatic mutations, and chimeric organisms.
3 This provides a novel way to pursue the idea that genes are a special kind of difference maker (Griffiths et al. 2015; Weber 2016).

## References

Backus, J. (1978). Can programming be liberated from the von Neumann style? A functional style and its algebra of programs. *Communications of the ACM*, 21(8), 613–641.
Bedau, M. A. (1997). Weak emergence. *Noûs*, 31(Suppl 11), 375–399.
Boudry, M., & Pigliucci, M. (2013). The mismeasure of machine: Synthetic biology and the trouble with engineering metaphors. Studies in history and philosophy of science Part C. *Studies in History and Philosophy of Biological and Biomedical Sciences*, 44(4), 660–668.
Calcott, B. (2014). Engineering and evolvability. *Biology & Philosophy*, 29(3), 293–313.

Griffiths, P. E. (2001). Genetic information: A metaphor in search of a theory. *Philosophy of Science*, 394–412.

Griffiths, P. E., Pocheville, A., Calcott, B., Stotz, K., Kim, H., & Knight, R. (2015). Measuring causal specificity. *Philosophy of Science*, 82(4), 529–555.

Jaeger, J., Laubichler, M., & Callebaut, W. (2015). The comet cometh: Evolving developmental systems. *Biological Theory*, 10(1), 36–49.

Keller, E. F. (2001). Beyond the gene but beneath the skin. In S. Oyama, P. E. Griffiths, & R. D. Gray (eds.), *Cycles of Contingency: Developmental Systems and Evolution*, 299–312. Boston: MIT Press.

Keller, E. F. (2003). *Making Sense of Life: Explaining Biological Development with Models, Metaphors, and Machines*. Cambridge, MA: Harvard University Press.

Malewicz, G., Austern, M. H., Bik, A. J. C., Dehnert, J. C., Horn, I., Leiser, N., & Czajkowski, G. (2010). Pregel: A system for large-scale graph processing. In *Proceedings of the 2010 ACM SIGMOD International Conference on Management of Data*, 135–146. SIGMOD '10. New York, ACM.

Moczek, A. P., Sears, K. E., Stollewerk, A., Wittkopp, P. J., Diggle, P., Dworkin, I., Ledon-Rettig, C. et al. (2015). The significance and scope of evolutionary developmental biology: A vision for the 21st century. *Evolution & Development*, 17(3), 198–219.

Newman, S. A., & Bhat, R. (2009). Dynamical patterning modules: A 'Pattern Language' for development and evolution of multicellular form. *The International Journal of Developmental Biology*, 53(5–6), 693–705.

Nicholson, D. J. (2014, December). The machine conception of the organism in development and evolution: A critical analysis. *Studies in History and Philosophy of Science Part C: Studies in History and Philosophy of Biological and Biomedical Sciences*, 48, 162–174.

Nijhout, H. F. (1990). Problems and paradigms: Metaphors and the role of genes in development. *BioEssays*, 12(9), 441–446.

Oyama, S., Gray, R. D., & Griffiths, P. E. (eds.). (2003). *Cycles of Contingency: Developmental Systems and Evolution*. Reprint ed. Cambridge, MA: A Bradford Book.

Peter, I. S., & Davidson, E. H. (2013). Pattern formation in Sea Urchin Endo mesoderm as instructed by gene regulatory network topologies. In V. Capasso, M. Gromov, A. Harel-Bellan, N. Morozova, & L. L. Pritchard (eds.), *Pattern Formation in Morphogenesis*, 75–92. Berlin, Heidelberg: Springer Proceedings in Mathematics.

Pigliucci, M. (2010). Genotype–phenotype mapping and the end of the 'genes as blueprint' metaphor. *Philosophical Transactions of the Royal Society of London B: Biological Sciences*, 365(1540), 557–566.

Planer, R. J. (2014). Replacement of the 'genetic program' program. *Biology & Philosophy*, 29(1), 33–53.

Vasarhelyi, G., Viragh, C., Somorjai, G., Tarcai, N., Szorenyi, T., Nepusz, T., & Vicsek, T. (2014). Outdoor flocking and formation flight with autonomous aerial robots. In *2014 IEEE/RSJ International Conference on Intelligent Robots and Systems*, 3866–3873. Chicago, IEEE.

Weber, M. (2016). Discussion note: Which kind of causal specificity matters biologically. *Philosophy of Science*, 84.

# 4 Living machines

## The extent and limits of the machine metaphor

*William Bechtel*

## Introduction

Machine metaphors, which interpret components of organisms as like machines humans have invented, have figured prominently in the life sciences since the 18th century. The "new mechanists" in contemporary philosophy of science (Bechtel and Richardson 1993/2010; Bechtel and Abrahamsen 2005; Machamer et al. 2000; Craver and Darden 2013; Glennan 2017) have borne witness to the tendency in many fields of biology to explain living phenomena as resulting, as with human-made machines, from the organized activities of the entities that constitute them. But like any metaphor, this analogy has its limits. Emphasizing entities or parts, activities or operations, and organization fails to bring out both the most potent insights and the limitations of the machine metaphor. Accordingly, I offer a different perspective.

In a series of papers starting in the 1960s, physicist turned theoretical biologist Howard Pattee advanced a framework for thinking about both the similarities and the differences between biological mechanisms and human-made machines[1] (several of the most important of Pattee's papers have been collected in Pattee and Rączaszek-Leonardi 2012). Unfortunately, this framework has been neglected in the philosophical literature on mechanisms. My goal in this chapter is to articulate three of Pattee's central insights about machines. Although all three apply to machines and biological mechanisms, important differences arise with respect to the third. First, machines and biological mechanisms are macroscale objects and as such cannot be characterized solely in terms of dynamical laws but require consideration of constraints. Second, the constraints in machines and biological mechanisms serve to enable these systems to perform work, which results from constraining flows of free energy. Third, for the work performed to be useful to the larger system in which the machine or mechanism is contained,[2] control mechanisms must operate on flexible constraints within the production machines or mechanisms. I use "production" to characterize those machines whose work generates products valued by the human user (e.g., a drill that produces a hole) or mechanisms whose work generates a product used by the organism (e.g., a ribosome that synthesizes proteins). Control mechanisms operate on production mechanisms, affecting whether or when they produce a product or the character of the product they produce.

As noted, it is with control that differences between human-made machines and biological mechanisms become manifest. In machines, control mechanisms are organized hierarchically so that at the highest levels the beneficiaries of the work, typically human beings, can determine what work is performed. In the case of biological mechanisms, while there is a sense in which the control mechanism is at a higher level than the production mechanism it controls, so that the local relation is hierarchical, the overall organization is not hierarchical but heterarchical. The beneficiary of the work performed by production mechanisms is the organism, which uses the work to maintain (or reproduce) itself. Hence, control resides primarily within the organism (although some organisms may allow more inclusive systems to perform some control operations on them). However, overall there is not a hierarchy of controllers with a highest-level controller, but a network of heterarchically organized control mechanisms, each of which carries out specific control processes. The coordination of the operation of production mechanisms needed for the organism to maintain itself is achieved through the interaction of control mechanisms, not through the action of a central executive.

A further reason control is so important for biological mechanisms is that they are so constrained that, when free energy is available, they carry out activities. We tend to think of machines as passive unless turned on. But many machines are designed to carry out activities unless deprived of a source of free energy. If one forgets to turn off a battery-powered appliance, it continues to run until the battery is depleted of power. Design that is conducive to activity is even more evident in biological organisms. Observe a bacterium or a squirrel. Even without stimulation, they are almost constantly moving. Sleep, a state (found in all animals in which it has been investigated) in which most motor and cognitive activities cease, exemplifies the perspective that biological mechanisms are endogenously active. Sleep requires control mechanisms that suppress mechanisms that would otherwise carry out their activities. Moreover, these mechanisms suppress only motor and cognitive activities – basic metabolic mechanisms continue to function apace. Control is needed not to initiate activity but to determine which production mechanisms are downregulated on a given occasion. This is especially clear when, as is often the case, different production mechanisms perform incompatible activities. In such cases, control mechanisms halt some production mechanisms while allowing others to function and then switch to allow the others to function.

The linchpin of Pattee's account is the notion of constraints. Constraints figure differently in explanations than do dynamical laws, leading Pattee to characterize a physicist as speaking a different language when talking of constraints than when talking of dynamics and of the need to recognize an epistemic cut between the two accounts (that is, the theorist applies a different theoretical framework in the two accounts). The next section, therefore, develops the framework of constraints while the section "Energy and work" considers how constraints enable production machines and mechanisms to perform work. In the section "Flexible constraints, information, and control," I turn to how production mechanisms are controlled, which requires developing further epistemic cuts and developing still other languages. First, control requires a specific kind of constraint in the production mechanism – one that is flexible and capable of being changed through the work performed by control mechanisms. Second, for the work done on flexible constraints to be effective in enabling machines to serve the ends of their human users or biological mechanisms to contribute to the self-maintenance of the organism, this work must be responsive to the conditions that require work. Accordingly, control mechanisms traffic in information. The section "Distinctive features of control of biological mechanisms" then shows how, in terms of this framework, we can understand the differences between machines and biological mechanisms: control of biological mechanisms depends not on an external user but on a network of control mechanisms within the organism.

## Dynamics and constraints

The crucial concept of "constraint" has its roots in classical dynamics (Sklar 2013; Hooker 2013), a field that faced the challenge of explaining the behavior of macroscale objects (rocks, organisms, etc.) in light of the fundamental force laws introduced by Newton. Force laws apply to all six degrees of freedom (three directional and three rotational) of each individual particle. One might undertake to account for the behavior of a macroscale object by determining the change in value along each degree of freedom of each particle that constitutes the macroscale object. If one could carry out the calculations from some initial condition, as Laplace envisioned, it would provide a complete account of the events in the universe. This would certainly be a Herculean task, but it is also unnecessary since within macroscale objects many degrees of freedom for constituent particles are constrained – either fixed, limited, or forced to change in conjunction with those of other particles. Moreover, focusing on individual particles ignores that which makes a collection of particles into a macroscale object. Macroscale objects result from the existence of physical and chemical bonds between particles that constrain the freedom of each particle. The particles that are bound together are now constrained to move in a coordinated fashion. Thus, when a force is applied to a macroscale object such as a table, all the particles within it are displaced together in the same direction.

Essentially, from the perspective of describing individual particles in terms of force laws, macroscopic objects are invisible. It is only by taking constraints into account that we identify macroscale objects. Since constraints freeze out changes in some degrees of freedom, these degrees of freedom can be ignored in accounts of macroscale objects. In physical accounts, constraints are represented in a different vocabulary than the dynamics. Constraints are not specified in terms of force laws, nor are they derived from them. Bonds that serve as constraints must be identified empirically. This is the context in which Pattee first introduces the idea of two languages:

> Whenever a physicist adds an equation of constraint to the equations of motion, he is really writing in two languages at the same time, although they may appear indistinguishable in his equations. The equation-of-motion

language relates the detailed trajectory or state of the system to dynamical time, whereas the constraint equation is not about the same type of system at all, but another situation in which some dynamical detail has been purposely ignored, and in which the equation of motion language would be useless.

(Pattee 1972)

In philosophical accounts of science that emphasize laws and view explanation as involving derivations from laws, the constraints appear as boundary conditions that must be added into the derivation (along with the initial conditions of the particles) to account for any specific phenomenon (Nagel 1961; Hempel 1965). The point to which Pattee is drawing attention is that boundary conditions are already foreign to the vocabulary of dynamical laws. In this section, I focus only on constraints that are treated as fixed over the time in which the changes of interest are occurring. In the case of fixed constraints, the challenge of integration with dynamical laws is relatively straightforward: constraints can be represented as parameters in the statements of the laws. From specifications of boundary conditions and force equations, one can often derive equations that characterize the behavior of specific constrained systems. A standard example is deriving the behavior of a pendulum from the force laws given boundary conditions about how a weight is connected to a pivot point. These constraints are referred to as "holonomic," as they can be described in terms of the coordinates of the particles constituting the system (and sometimes time). Even when they can be integrated into a common account, it is important to recognize that constraints are different from forces. This is a difference that obtains greater significance when we turn to flexible constraints.

The language of constraints is an alternative to describing machines or mechanisms in terms of the activities or operations performed by constituent entities or parts. The value of this alternative characterization will be developed in subsequent sections. Here I simply draw the connection to the more traditional vocabulary of activities or operations. The organization of parts constrains them to behave in a coordinated manner. Take a simple machine such as a lever, which consists of a fixed beam and a hinge or fulcrum. The particles in the fixed beam are constrained to move together. As a result, once a force is applied to one end of the beam, it leverages the movement of the weight on the other end. The relation of the fulcrum to the beam serves as a constraint that ensures that as one end is depressed, the other end moves up and raises the weight. More complex machines such as clocks combine multiple parts (cogwheels, etc.) that transfer the movement of one part (unwinding of a spring) into the movement of other parts (the hands of the clock) in a regular manner. Membranes and enzymes are two examples of constraints in living organisms. Membranes provide selective barriers that constrain some molecules to stay on one side, separated from those on the other side. Enzymes are molecules that bind to other molecules, substrates, which they constrain to be in appropriate positions to either form or break chemical bonds. Enzymes are often constrained within membranes, which serve such ends as keeping one part of the enzyme in a hydrophilic environment and the other in a hydrophobic environment. The combined constraint of an enzyme situated in a membrane is able to accomplish such things as moving selected molecules across the membrane.

In these examples of machines and biological mechanisms, the parts and organization impose constraints and thereby determine what activities the mechanism as a whole will perform as a result of the activities of each of its parts.[3] Viewing machines and mechanisms in this way also makes clear a point Hooker (2011) has emphasized: constraints are not just limiting but also enabling. As a result of limiting degrees of freedom of individual particles, all the constituents can be altered in a coordinated fashion and thereby reach locations they cannot otherwise reach. For example, water spilling on a surface normally spreads in all directions, but in a pipe it is limited to moving in one direction, resulting in an accumulation at the end of the pipe. An illustrative example Hooker provides is that a skeleton limits the freedom of motion of muscles in one's body. But it also so fixes the position of muscles that organisms with skeletons can do things that are not possible in organisms lacking a skeleton.

In this section, I have introduced constraints as giving rise to macroscale objects by limiting degrees of freedom of constrained components. Among the macroscale objects made possible by constraints are machines and biological mechanisms. Machines and mechanisms are constrained systems that, as a result of being constrained, are able to do things that their constituents alone would not be able to do. The value of the constraint perspective will begin to become apparent in the next section. In the sections that follow, I will show more clearly the importance of differentiating the language used to describe constraints from that in which the dynamics is characterized in terms of force laws.

## Energy and work

Construing machines and biological mechanisms in terms of constraints begins to generate new insights when we consider that what is constrained is the flow of free energy. Constraining the dissipation of free energy enables the performance of work. "Energy" and "work" are concepts that have not been employed in philosophical discussions of biological mechanisms, but what mechanisms do is change the state of the world, and this requires the constrained dissipation of free energy. Focusing on energy, as physicists do in thermodynamics, also brings a shift in how events in the world are described. Newton's laws are reversible: given what is specified in the laws, the temporal order of any sequence of events could be reversed. But in fact they occur in a particular order. What determines the order in which events actually occur is the flow of free energy, which results from the unequal distribution of particles in the universe. Starting from a specific, non-equilibrium distribution, particles move in a manner that results in their being, statistically, more equally distributed. In the language of thermodynamics, overall entropy increases.

This overall pattern of moving from ordered low-entropy or high free-energy states to less ordered states does not itself result in work. Work results when the flow of free energy is constrained to generate coherent patterns. A pipe, as noted earlier, can constrain water to move in limited ways. When a pipe on Earth is oriented downward and there is a source of water at the top, then the free energy available in the water given its relative height results in a force at the bottom that can be used to perform work. The work may just serve to carve out a gulley in the terrain at the bottom opening of the pipe, but it can also be captured and used in a milling machine to rotate a grinding wheel that performs the work of grinding grain. Steam engines work by similar principles: combustion liberates the available energy in the fuel to create steam, which is constrained to flow through pipes until it is used to operate other machines such as sewing machines.

Biological mechanisms are no different than human-built machines in this respect – both require sources of free energy to perform work. The one notable difference is that the work biological mechanisms perform is mostly directed to building and repairing the mechanisms that constitute the organism. In plants, radiation from the sun provides the source of free energy, which is then constrained in a sequence of chemical reactions to synthesize molecules such as glucose. The bonds in glucose and other organic molecules themselves can provide free energy, which plants as well as other organisms that eat plants can use to synthesize other molecules or carry out other activities such as locomotion (which organisms use in part to procure additional free energy). All work in organisms requires free energy, and an important perspective to take on mechanisms is that they provide constraints that direct flows of free energy into the performance of work.

Philosophical accounts of biological mechanisms largely ignore the role of free energy, as does much biological research on mechanisms. For example, the mechanism for protein synthesis at the ribosome is viewed as adding amino acids to a sequence in an order that corresponds to the order of nucleic acids found in mRNA. In analyzing the history of research leading to the current account of protein synthesis, Darden and Craver (2002) follow the lead of scientists who emphasize the construction of a sequence of amino acids that corresponds to the sequence of nucleic acids in mRNA, resulting in the primary structure of the protein. Yet, as they note, one of the early research projects attempting to explain protein synthesis focused on the energy required. Clearly, both a source of free energy and sequence information are critical to create primary structure. It is important to recognize that they play very different roles in the mechanism. The sequence information coded in mRNA is a relatively static constraint in the synthesis process (new mRNA sequences are ferried to the ribosome but at a sufficiently slow rate such that, during the time of transcription of a single protein, the mRNA does not change). The activity of a tRNA bearing an amino acid docking to a codon on the mRNA, adding a bond between the amino acid it ferries and the amino acid sequence constructed so far, only occurs because free energy, in the form of GTP, is constrained to perform the work required at two critical steps: accommodation and translocation (Figure 4.1). Without the free energy, the mRNA would not be translated into the primary sequence of a protein.

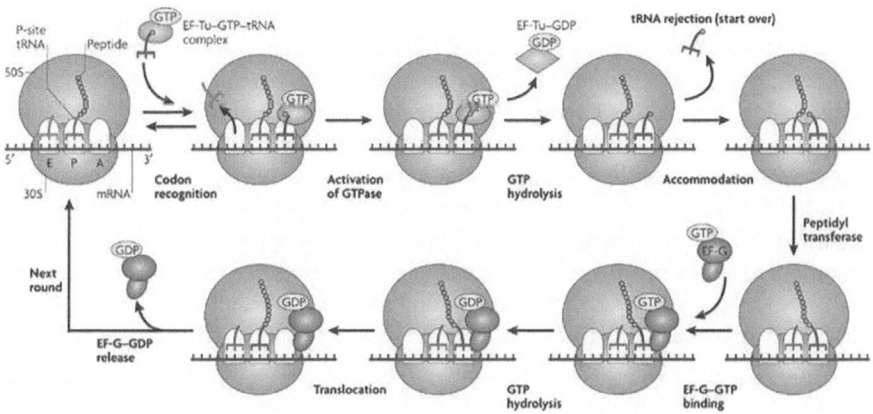

Nature Reviews | Molecular Cell Biology

*Figure 4.1* Hydrolysis of GTP to release energy figures in two crucial steps in protein synthesis. In the reactions shown on the top row, the incoming tRNA is bound to (EF)-Tu–GTP. If the codon-anticodon pairing is correct, this activates the GTPase in the ribosome, which initiates hydrolysis of GTP to provide the free energy to add the amino acid to the forming chain. GTP hydrolysis also provides the energy to translocate the machinery along the mRNA.

Source: From Steitz, 2008.

Note: This figure can be accessed in color via the eBook version of the book and eResources at www. routledge.com/9780815380788.

In this section, I have introduced the important role constraints play in directing the flow of free energy to enable the performance of work such as that involved in protein synthesis. Without a source of free energy, neither machines nor biological mechanisms would do anything other than gradually degrade as part of the inevitable increase of entropy. Channeling free energy enables the performance of work, which can result in building structures, such as proteins, that are at a lower entropy state, albeit resulting in an overall increase in entropy in the larger system. For the capacity to perform work to be put to use either for human ends or for maintaining a living organism, it must be directed in the needed way. This requires control.

## Flexible constraints, information, and control

Control of a given mechanism is achieved by another mechanism operating on it. A control mechanism consists of two components. One alters the operation of the production mechanism by acting on a flexible constraint. The second detects and represents information either about the state of the production machine or mechanism itself or conditions in its environment. Each component generates an epistemic cut beyond that provided by the introduction of constraints alone.

If a machine or mechanism contained only fixed constraints, it would perform work until the supply of free energy was exhausted. If some of the constraints are flexible, then they can be altered. To stop a machine from continuing to run until its energy is depleted, engineers employ switches that either directly or indirectly alter constraints within the machine. The ignition switch in a traditional automobile allows drivers to start or stop the car's engine from operating by changing a constraint in the circuitry between the battery and the spark plugs. Switches can have more than one setting, and some flexible constraints can take continuous values. For example, in addition to constraining the engine to run or not, by controlling gears a driver can determine whether the car moves forward or backward, and by depressing the accelerator a specific amount the driver determines its speed. In each case, there is a linkage that transmits (in some cases enhances) the work performed by the driver in altering the controls to the flexible constraints within the production mechanism. Thus, in cars with a carburetor, the driver's action with respect to the accelerator pedal determines the opening of a value that lets fuel into the carburetor and then into the engine.

Flexible constraints seriously complicate the ability to represent a mechanism mathematically. They cannot be treated as fixed boundary conditions. Rather, the account of how the mechanism operates must take into account how the constraints change. In a mathematical analysis, the constraint can no longer be specified just in terms of positions and relations between the components within the production machine or mechanism. Such constraints are non-holonomic. To date, physicists have not developed generalized ways to incorporate non-holonomic constraints into the basic force laws as presented in either Hamiltonian or Lagrangian formalisms (Hooker 2011, 2013). Rather, researchers must combine two modes of analysis. Pattee's point about the need for two languages now assumes much greater import. There is an epistemic cut between the account of how the values of flexible constraints are determined and the account of the dynamics of the constrained system, and at present there is no real hope of unifying these into one common account.

A further epistemic cut arises when we turn to the information upon which the activity of the control mechanism is based. The control mechanism is itself a constrained dynamical system, and one could describe its operation in much the same manner as the production machine or mechanism. For example, one might characterize the operation performed on the valve in the carburetor as a constrained dynamical system. This can be accomplished relatively easily if all one considers is the linkage mechanisms between the accelerator pedal and the carburetor, but that ignores a critical part of the control system: the human that is depressing the accelerator (or the system that performs this function in a self-driving car). A far more challenging task would be to explain all the relevant activities within the human that resulted in depressing the accelerator pedal. Even if such an account were forthcoming, it would not provide the relevant description for understanding control. Part of the basis for the driver's action is her perception of road conditions. The driver is detecting such things as how quickly she is approaching the car in front of her. (Other considerations, such as the driver's desire to get to her destination quickly, also affect what control she exercises, but I will not pursue those here.) We characterize the driver as representing relevant information and basing control actions on her representations. In characterizing the driver in this way, we are taking a very different epistemic perspective than we take in describing her as simply a dynamical system.

The same type of cut between information processing and dynamics is required to understand control systems that do not involve cognitive agents such as humans. The governor Watt designed for the steam engine provides a well-studied example. Watt's objective was for the production machine, the steam engine, to operate at the same rate regardless of the resistance supplied by the machinery attached to the engine. He did not want the rate to increase if, for example, several operators of sewing machines attached to it took a rest break. Watt targeted a valve in the steam supply as the flexible constraint on which his governor would work and then designed a control mechanism that would measure the change in velocity of the steam engine due to changing resistance from the downstream machinery and represent this information in a format that could be used to direct action on the valve. To detect the change in velocity and represent it, Watt connected a spindle to the flywheel that would rotate at the same velocity, and he attached flexible arms to it that would, as a result of centrifugal force, extend further out when resistance dropped and drop down when resistance increased. One can, as Maxwell (1868) did, analyze the behavior of the governor in a set of differential equations that characterize its operation. These are distinct from the equations that describe the dynamical behavior of the engine itself. More importantly, they do not specifically identify the information relation – that the angle arms represent information about the change in speed of the flywheel.[4] This relation is not privileged in the dynamical account, but it is the relation that is crucial to its control function. One way to appreciate this point is to note that it is this information relation that is constant across all implementations of the Watt governor even as the dynamics vary depending on the specific materials used in the construction of the governor.

Watt's governor is one example of a design for control known as "negative feedback" – departures from a target operation of a machine are detected and represented and the representation is used to alter the operation of the production machine. Once negative feedback was recognized as a general design principle for maintaining a measured quantity at a constant value, it was applied not only in the design of many other machines (e.g., heat-seeking missiles) but also in the attempt to understand how biological organisms maintain what Bernard (1865) referred to as the "constancy of the internal environment." This requires that each of the production mechanisms that performs work that changes the internal environment has a flexible constraint and that this constraint be altered by the control mechanism as it measures and represents relevant information.

These two roles – detecting and representing information and acting on flexible constraints – are realized in allosteric enzymes, which contain two sites at which substrates can bind. At one site the enzyme binds to a substrate and catalyzes the reaction. At the other site the enzyme binds to a signaling molecule, which carries information. The effect of the enzyme binding the signaling molecule is to change the conformation of the enzyme and alter the ability of the catalytic site to catalyze a reaction. It is thereby altering a flexible constraint. To make the difference in these two roles clear, it helps to think of the enzyme as consisting of two different components.

The usefulness of differentiating the two parts of an allosteric enzyme is illustrated by phosphofructokinase, which catalyzes a reaction early in glycolysis. At both sites the enzyme binds to ATP, but the binding at each site plays a different role. When ATP binds to the catalytic site, the enzyme catalyzes a reaction that transfers a phosphate group from ATP to fructose-6-phosphate to create fructose-1,6-diphosphate. In this role, ATP is simply the source of a phosphate group that becomes bound to the substrate. Even though the effector site also binds ATP, the consequences are very different. The effect is to change the conformation of the enzyme in such a manner that the catalytic site is no longer active.

The effector site and the process of changing the enzyme's conformation constitute a second control mechanism. ATP constitutes the free energy that is available to the cell to perform work (not just provide phosphate bonds to fructose-6-phospate). If the concentration of ATP is sufficiently high, the cell does not need more to be produced, and allowing more glucose to pass down the glycolytic pathway is counterproductive.

By measuring the concentration of ATP through its ability to bind to the effector site and using that information to alter the catalytic site, part of the enzyme performs control over another part. By measuring ATP concentrations and using those to determine the conformation of the enzyme, phosphofructokinase functions as a control mechanism operating on the glycolytic mechanism.[5]

As both the Watt governor and the glycolytic mechanism make apparent, control requires linking together a measurement or detection system that identifies conditions in which different activities are required with a means to operate on flexible constraints in the production machine or mechanism. Understanding control thus requires integrating an account of how information is procured and represented with an account of the mechanism whose dynamics will be altered. The flexible constraint is the point of interface at which two languages, one describing the dynamics of the constrained system and the other how information is used to change the constraint. For a complete account, these must be integrated. There is, however, in general no systematic way to integrate the two representations. Rather, to understand control as Pattee characterizes it, scientists must speak in both languages, switching between them as needed. One cannot be reduced to the other.

## Distinctive features of control of biological mechanisms

So far I have treated human-built machines and biological mechanisms as comparable. The only significant difference I have noted is that the work performed by biological mechanisms is generally directed at making or repairing the mechanisms constituting the organism, not for the benefit of an external user. This difference, though, has corresponding implications for how machines and biological mechanisms are controlled. Control in machines is needed to ensure that the machine performs the specific work desired by human users. Control tops out with the human user. Control of biological mechanisms is not primarily in the service of an external agent (although sometimes external agents – other organisms or social collectives – can exercise control over them) but is in the service of the collection of mechanisms constituting the organism itself. Control serves to enable mechanisms to perform the work needed for the whole collection of mechanisms constituting an organism to survive (or reproduce).

This difference is seen clearly if we focus on repair.[6] Both machines and biological mechanisms are far-from-equilibrium systems that break down over time as they move toward equilibrium. Humans build machines out of relatively long enduring materials such as wood and metal that degrade relatively slowly. When a component of a machine breaks, humans intervene to either repair the component or replace it (or, increasingly, replace the whole machine). In both respects, biological organisms are different. Being constituted of more transient chemical systems, transitions toward higher entropy occur more frequently in organisms than in machines. This requires nearly constant inspection and repair of biological mechanisms. In addition, rather than relying on an external agent, repair of biological mechanisms must be carried out by the collection of mechanisms constituting the organism.

In fields like cell biology, researchers were slower to identify control mechanisms than production mechanisms since production mechanisms are more readily detectable. They can be detected by the phenomena they produce, and experimental procedures are designed to remove control so as to generate these phenomena reliably. Research on diseased cells such as cancer cells has provided avenues to investigate control mechanisms. Cancer cells employ the same production mechanisms as healthy cells, but these mechanisms operate differently due to altered control mechanisms. For example, cancer cells continually enter into the cell cycle and divide, whereas most mature healthy cells in biological tissues stop dividing. Analyzing how various somatic mutations figure in altering control mechanisms in cancer has provided insight into just how complex and prevalent these control mechanisms are and the effects they have on production mechanisms (Hanahan and Weinberg 2000, 2011; Bechtel 2018). In Figure 4.2, some of the proteins are coded for by major oncogenes (e.g., Ras and PTEN) and tumor suppressor genes (e.g., Rb and p53). These are situated in signaling pathways (control mechanisms) that end up controlling the expression of many genes that figure in numerous cell mechanisms.

Given the number of control mechanisms in individual organisms, a natural question is, how are they organized so as to support the ability of organisms to maintain themselves? One could easily imagine different control mechanisms generating conflicting activities in the production mechanism. This is a question for which we do not yet have a full answer. But we can make some progress by considering how control mechanisms are organized.

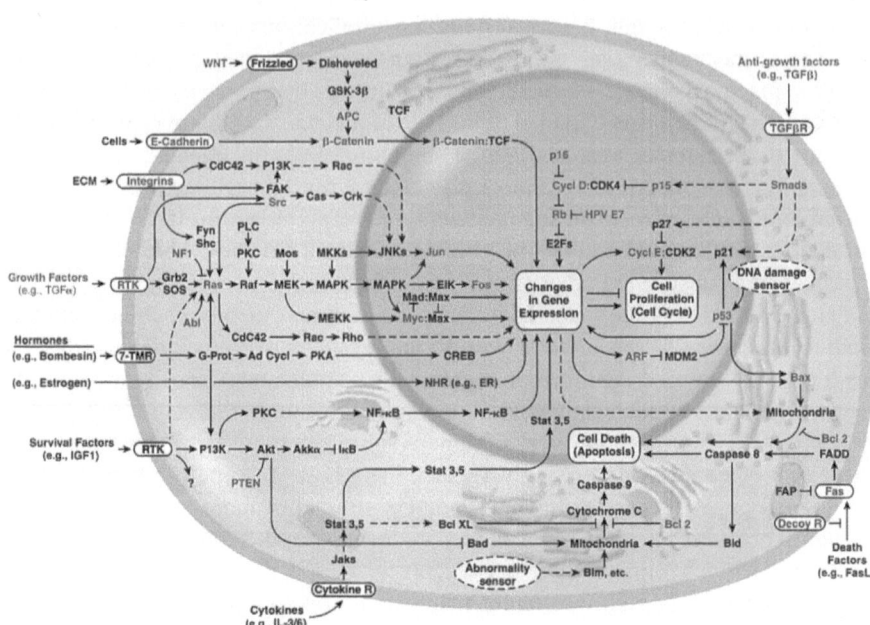

*Figure 4.2* Major pathways through which oncogenes and tumor suppressor genes, shown in red, figure in signaling pathways that control various cell mechanisms.

Source: From Hanahan and Weinberg 2000.

Note: This figure can be accessed in color via the eBook version of the book and eResources at www.routledge.com/9780815380788.

Control is naturally understood as hierarchical: a control mechanism operates on a production mechanism. Moreover, a control mechanism can operate on a flexible constraint within another control mechanism, generating a multi-tiered hierarchy of control mechanisms. Nervous systems are often interpreted in such hierarchical terms. Figure 4.2, however, suggests a different picture. First, most of the signaling pathways begin with a receptor on the cell surface that detects a signal from the environment, such as growth factors, hormones, and cytokines. Others initiate with detectors of internal conditions, as indicated by the two dashed ovals. Second, the various signaling pathways are also interconnected at a variety of points. This suggests that rather than being assembled into a hierarchy with a highest-level controller at the top, control mechanisms are organized within a network. Although we have little understanding of how various controllers are coordinated through such a network, it is presumably through the interactions within the network that the various control mechanisms coordinate their activities so as to engage and disengage production mechanisms in ways that maintain the whole system of production and control mechanisms.

As I have noted, the nervous system is usually interpreted as a hierarchical system, with the brain controlling lower-level controllers and the neocortex controlling mechanisms in lower brain areas. In what follows, I take a few steps toward presenting a different picture. Neural control mechanisms often exercise control over muscles – complex, constrained mechanisms that are capable of contracting. The core of a muscle mechanism consists of two complex molecules (already highly constrained systems): actin and myosin. Myosin constrains the release of free energy (in the form of ATP), directing it into the construction of crossbridges between pairs of actin and myosin molecules (adopting a fibril shape). These crossbridges cause the two molecules to slide along each other. If nothing else operates, muscle will contract whenever energy is available. To prevent that from happening, tropomyosin blocks the binding between actin and myosin. When bound to calcium, however, tropomyosin no longer performs this function. It is thus a flexible constraint: when calcium is released in the cell, tropomyosin no longer interferes with the creation of bridges between actin and myosin. Calcium is stored in the sarcoplasmic reticulum, from which it is released when the binding of neurotransmitters to receptors on the cell surface generates electrical current that acts on another, higher-level, flexible constraint.

Although this account is highly simplified, it presents a picture of an intracellular hierarchy of controllers operating on other controllers that ultimately affects whether a muscle contracts. A common way to extend this account is to trace a neural pathway up to the highest brain centers that exercise control over muscle contraction. Keijzer et al. (2013), however, advance a different perspective on how neurons function as a control system for muscles: the "brain-skin thesis." According to this theory, the function for which neurons first evolved was to enable different muscles to coordinate their contractions. Keijzer et al. use jellyfish to illustrate this thesis. Within the skin (indicated as epidermis in Figure 4.3) of the bell of the jellyfish are muscle fibers. The jellyfish is propelled forward by coordinated contraction of these muscles, which pushes water out of the bell. Coordination is achieved through a nerve net overlaying the muscles. What is distinctive about this account is that rather than implementing higher-level decisions to contract muscles, this network transmits signals between muscles. This allows the individual muscles to effect control over each other, enabling nearby muscles to synchronize their contractions.

In addition to the interaction in the nerve network, coordinated activity is further supported by the inner of the two nerve rings, which, as shown in Figure 4.3, surround the bell of the jellyfish. The inner ring neuron functions as a pacemaker, generating a rhythmic contraction approximately every two seconds. The nerve net and the inner nerve ring constitute control mechanisms operating on muscles that enable the jellyfish to maintain endogenous activity. Unless interfered with, they enable the jellyfish to generate an ongoing pattern of swimming forward and then trolling for food.

While this primary control mechanism enables endogenous activity, the jellyfish is also able to respond to conditions in its environment, such as the presence of predators. This is accomplished by a further control mechanism that detects

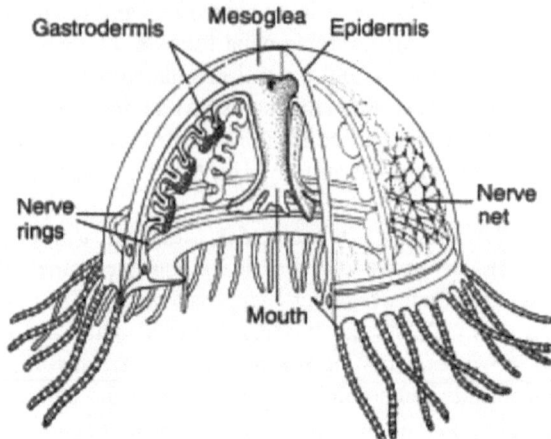

*Figure 4.3* Major anatomical elements in the jellyfish. Of particular importance are the nerve net, just inside the epidermis of the bell of the jellyfish, and the two nerve rings, which go around the bell.

perturbations to the tentacles and sends a signal around the bottom of the bell of the jellyfish using the outer nerve ring. This signal serves to alter flexible constraints within the nerve net so as to generate a much more intense swimming activity that allows the jellyfish to escape predators. This control system operates on top of the other control systems just discussed, perturbing their operation.

The pattern of organization exhibited in the jellyfish is widespread. Muscles are mechanisms primed to contract when acted on by a neuronal control mechanism. Pattern generators are key elements of neuronal control systems in many species. They create the potential for endogenous patterns of muscle contraction even in the absence of other signals. But these control mechanisms employ flexible constraints that can be acted on by other control mechanisms, such as ones that detect predators or resource availability. As in the jellyfish, these controllers can be highly specialized, serving to produce a specific activity when activated. Because they act on the pacemaker system that integrates the multiple muscles into a coordinated system, these specialized control systems can themselves elicit coordinated activity.

These networks can serve to integrate multiple signals and even allow the organism to decide between competing courses of action. For example, in each segment of the leech there is a network of approximately 400 neurons that, among other things, coordinates the muscle activities required for two different motor behaviors: crawling and swimming. Using a paradigm in which they present stimuli that are neutral between eliciting a swim or crawl response, Briggman et al. (2005) show how a decision to swim or crawl emerges dynamically through the interactions in the local network, involving in part neurons that constitute a pattern generator. As in the jellyfish, a network of neurons provides an integrative system of control mechanisms in which a coordinated response can emerge.

The perspective that emerges from both the intercellular signaling network revealed in cancer and the simple nervous systems in the jellyfish and leech is that control of biological mechanisms is not strictly hierarchical and is often mediated by networks in which multiple control processes can be integrated. This is what might be expected if many individual control mechanisms each detect specific conditions and alter specific flexible constraints in other mechanisms. No individual control mechanism exercises overall control, as the user of a human-built machine does. In the evolution of the organism, such control mechanisms developed opportunistically: when a core mechanism has a flexible constraint and another mechanism develops that can detect information and use it to alter the flexible constraint in a manner that, from the perspective of the whole organism, improves, or at least doesn't seriously impair, the functioning of the whole organism, then it may be added into the network.

A convoluted network of control mechanisms is appropriate for biological organisms that, for the most part, must rely on themselves to construct and repair the constraints that constitute the collection of mechanisms they comprise. While open to energetic and material inputs from the environment, organisms carry out all the activities that construct and repair their own mechanisms and take actions in the world that contribute to such self-maintenance. This is the feature of organisms that many theorists highlight by calling them "autonomous" (Varela 1979; Moreno and Mossio 2015). Autonomy does not mean that other organisms or conditions in their environment do not influence autonomous systems or even at times exercise control over them by performing work on flexible constraints within them. Individual cells can be incorporated into multicellular organisms and different species can form mutually beneficial collaborations (e.g., humans and the wide range of bacteria constituting our microbiome). In both cases, control mechanisms in the collective operate on flexible constraints in individual organisms. Even in these arrangements, the individual living organisms constrain the flow of free energy in ways that build and maintain themselves through complex, integrated control networks.

## Conclusion

To explore the scope and limits of applying the machine metaphor to biological mechanisms, I have followed Pattee in focusing on constraints in establishing the identity of macroscale objects. As in machines, constraints in biological mechanisms direct flows of free energy to perform work. In both machines and biological mechanisms, many of these constraints are fixed. Others are flexible, and it is flexible constraints that enable other machines or mechanisms to control production machines and mechanisms. This broad framework applies equally to machines and biological mechanisms.

Focusing on constraints yields a different perspective on machines and mechanisms than the new mechanists' focus on entities and activities. As I have noted, identifying constraints entails what Pattee characterized as an epistemic cut. Constraints are distinct from the dynamic processes they constrain: they cannot be derived from dynamical force laws but must be discovered empirically. Even when they are fixed and can be treated as holonomic, they are described differently from the force laws, and so Pattee introduced the idea of using different languages to characterize constraints and to describe dynamics. But when constraints are flexible, they create an even greater rupture in the fabric of science since currently there is no way to incorporate such non-holonomic constraints into the equations characterizing forces within the production machine or mechanism. Yet flexible constraints are crucial for allowing machines and mechanisms to be controlled by other mechanisms (including human operators in the case of machines). Control mechanisms are characterized in terms of information: they detect or measure conditions, represent them, and use those representations to alter constraints in production mechanisms. One can give a dynamical account of the operation of control mechanisms, but their role in carrying out control can be understood only in information terms: the control activity is based on the information that is detected. This epistemic cut is deep, but to understand controlled machines and mechanisms, investigators need to employ both perspectives: they must speak the language of both dynamics and information.

This framework of constraints and control also provides a perspective on the differences between machines and biological mechanisms. Machines perform work for their human users and are designed so that users ultimately control them. Much of the work of biological mechanisms is directed at the organism constituted by the collection of mechanisms – building and repairing the various mechanisms or replicating the whole system. This has broader implications when we turn to control mechanisms. Rather than relying on an external agent to collect information and make decisions, individual control mechanisms detect and represent information and use it to alter flexible constraints in production mechanisms. There is no agent overseeing the whole collection of mechanisms and making decisions. Since individual production mechanisms and their coupled control mechanisms may operate at cross-purposes, there is need for coordination so that within the organism the necessary activities of building and repairing mechanisms or replicating do not conflict with each other. This, I have suggested, is achieved though networks of control mechanisms. We know very little at the moment about how the networks of control mechanisms in organisms accomplish this feat. One part of the explanation is that such networks can create patterns of endogenous activity that more specialized control mechanisms can modify. We may hope in the future to discover design principles that describe networks of controllers that allow for biological autonomy. This is where we should look to understand more fully the differences between machines and biological mechanisms.

## Acknowledgments

I thank Jason Winning for many valuable discussions of the topics in this chapter and Leonardo Bich for helpful comments on a previous draft.

## Notes

1  Just as many human-made machines have been inspired by biological mechanisms, I leave open the possibility that humans may create machines that are sufficiently like biological mechanisms that the distinction will not apply. In this chapter, I use "machine" for the types of machines that humans have been building over the last several centuries.
2  In the case of biological mechanisms, the immediate larger system is the cell, while the larger system for machines may just include the machine and the human user but may also involve a social unit such as a factory.
3  In other work, Winning and Bechtel (2018) have argued that attention to constraints allows mechanists to solve a problem for which they otherwise lack a solution: explaining why parts of mechanisms have the causal powers they do. Appealing to constraints allows mechanists to advance beyond a Humean construal of causation to a richer notion of cause that supports counterfactual reasoning about mechanisms.
4  Once one has the information relation specified, one can, as Nielsen (2010) showed, solve Maxwell's equations to show how change in engine speed due to resistance is measured. Surprisingly, the representation of the information requires not just the location of the angle arms but also the velocity and acceleration of their movement.
5  There are limitations to feedback controllers. For example, they are limited to altering the behavior of the production mechanism on the basis of the output of the mechanism they control. Also, they are incapable of producing qualitatively different states (Bich et al. 2016). These limitations do not affect the primary point that negative feedback controllers employ information. Moreover, there are ready strategies for enhancing the processing in control systems. The controller may be engineered to respond to other information, such as conditions in the mechanism's environment. And design principles, such as double-negative feedback, can be implemented to enable switching between two different regimes.
6  In much of his work, Rosen (1991) emphasized the importance of repair in biological systems.

## References

Bechtel, W. (2018). The importance of constraints and control in biological mechanisms: Insights from cancer research. *Philosophy of Science*, 85, 573–593.

Bechtel, W., & Abrahamsen, A. (2005). Explanation: A mechanist alternative. *Studies in History and Philosophy of Biological and Biomedical Sciences*, 36, 421–441.

Bechtel, W., & Richardson, R. C. (1993/2010). *Discovering complexity: Decomposition and Localization as Strategies in Scientific Research*. Cambridge, MA: MIT Press. edition published by Princeton University Press.

Bernard, C. (1865). *An Introduction to the Study of Experimental Medicine*. New York: Dover.

Bich, L., Mossio, M., Ruiz-Mirazo, K., & Moreno, A. (2016). Biological regulation: Controlling the system from within. *Biology & Philosophy*, 31, 237–265.

Briggman, K. L., Abarbanel, H. D. I., & Kristan, W. B. (2005). Optical imaging of neuronal populations during decision-making. *Science*, 307, 896–901.

Craver, C. F., & Darden, L. (2013). *In Search of Mechanisms: Discoveries Across the Life Sciences*. Chicago: University of Chicago Press.

Darden, L., & Craver, C. (2002). Strategies in the interfield discovery of the mechanism of protein synthesis. *Studies in History and Philosophy of Biological and Biomedical Sciences*, 33, 1–28.

Glennan, S. (2017). *The New Mechanical Philosophy*. Oxford: Oxford University Press.

Hanahan, D., & Weinberg, R. A. (2000). The hallmarks of cancer. *Cell*, 100, 57–70.

Hanahan, D., & Weinberg, R. A. (2011). Hallmarks of cancer: The next generation. *Cell*, 144, 646–674.

Hempel, C. G. (1965). Aspects of scientific explanation. In C. G. Hempel (ed.), *Aspects of Scientific Explanation and Other Essays in the Philosophy of Science*, 331–496. New York: Macmillan.

Hooker, C. A. (2011). Introduction to philosophy of complex systems: A. Towards a framework for complex systems. In C. A. Hooker (ed.), *Handbook of Complex Systems: Handbook of the Philosophy of Science*, Vol. 10, 3–90. Amsterdam: Elsevier.

Hooker, C. A. (2013). On the import of constraints in complex dynamical systems. *Foundations of Science*, 18, 757–780.

Keijzer, F., van Duijn, M., & Lyon, P. (2013). What nervous systems do: Early evolution, input-output, and the skin brain thesis. *Adaptive Behavior*, 21, 67–85.

Machamer, P., Darden, L., & Craver, C. F. (2000). Thinking about mechanisms. *Philosophy of Science*, 67, 1–25.

Maxwell, J. C. (1868). On governors. *Proceedings of the Royal Society of London*, 16, 270–283.

Moreno, A., & Mossio, M. (2015). *Biological Autonomy: A Philosophical and Theoretical Inquiry*. Dordrecht: Springer.

Nagel, E. (1961). *The Structure of Science*. New York: Harcourt, Brace.

Nielsen, K. (2010). Representation and dynamics. *Philosophical Psychology*, 23, 759–773.

Pattee, H. H. (1972). The nature of hierarchical controls in living matter. In R. Rosen (ed.), *Foundations of Mathematical Biology, Vol. 1: Subcellular systems*, 1–22. New York: Academic Press.

Pattee, H. H., & Rączaszek-Leonardi, J. (2012). *Laws, Language and Life: Howard Pattee's Classic Papers on the Physics of Symbols with Contemporary Commentary by Howard Pattee and Joanna Raczaszek-Leonardi*. Dordrecht: Springer.

Rosen, R. (1991). *Life Itself: A Comprehensive Inquiry into the Nature, Origin, and Fabrication of Life*. New York: Columbia.

Sklar, L. (2013). *Philosophy and the Foundations of Dynamics*. Cambridge: Cambridge University Press.

Steitz, T. A. (2008). A structural understanding of the dynamic ribosome machine. *Cell Biology*, 9, 242–253.

Varela, F. J. (1979). *Principles of Biological Autonomy*. New York: North Holland.

Winning, J., & Bechtel, W. (2018). Rethinking causality in neural mechanisms: Constraints and control. *Minds and Machines*, 28(2), 287–310.

# Part 2
# Methodological issues

# 5   Beyond machine-like mechanisms

*Arnon Levy and William Bechtel*

## Introduction

Human-made machines of the industrial age, such as mills, typewriters, looms, and engines, provide vivid exemplars of a complex yet comprehensible type of system. As such, machines have served as an attractive paradigm around which to organize biological research, especially within proximate disciplines like cell biology, genetics, and physiology. For biologists, viewing an organism as a collection of little machines, a factory as it were, has provided both an image of what a successful explanation should look like and a set of methods and conceptual tools for investigation – exploring the parts and structures constituting the system, coming up with potential explanations, and determining whether they are correct. However, a science organized around the image of a machine[1] has its limits. Present-day biology appears increasingly to be approaching, and in many cases transgressing, these limits. Our goal in this chapter is to highlight both the contributions and the limitations of the machine image and to begin to indicate how a biology that moves beyond it, but still appeals to mechanisms, might look. Our focus is on how an investigation centered around such systems is pursued – which methods it employs and how such methods differ from those associated with the machine image. Thus, our focus is methodological. To be sure, there is a range of topics to which the status of the machine analogy is relevant – including more abstract issues like reduction and levels, theory structure, and the nature of explanation – but the topics we focus on have a more immediate, practice-based character, and we therefore see them as a good starting point.

We will devote most of our attention to four features of machines: they are (1) enduring structures that are readily differentiated from their environment that (2) consist of spatially localized and enduring parts, which (3) perform specific operations and (4) are organized so that whenever their start-up conditions are realized they produce the desired product. The overall functioning of such machines can usually be characterized qualitatively, emphasizing the effects parts have on each other and the order in which parts make their contribution to the production of the phenomenon.[2] These features are also found in the way biological mechanisms have been portrayed by the so-called new mechanists (Machamer et al. 2000; Bechtel and Abrahamsen 2005; Glennan and Illari 2018), and they are associated with theses they have put forward with regard to exploration, discovery, and confirmation.

The machine picture, at least to a first approximation, does fit a number of examples in biology, and the new mechanists have offered valuable analyses of these cases. However, each of these features is absent or attenuated in many other systems studied in contemporary biology. In many cases, the working parts are not entities but concentrations of entities (or an otherwise diffuse set of objects). Especially when the effects of the parts are nonadditive, describing the operation of these systems commonly requires quantitative, not just qualitative, accounts. Some of the systems are generated only as needed. Even those that are enduring often lack stable parts or are substantially reorganized as the system produces the phenomenon. Attending to such cases and comparing them to more machine-like cases not only allows us to see how the machine metaphor was historically conducive to discovery and understanding but also alerts us to the limits it imposes.

A few further comments before we proceed. What is the relation between our discussion and the mechanistic program in philosophy of science? And, more generally, how are machines related to mechanisms and how does a discussion of the limits of machine analogies contribute to our understanding of mechanistic science? We think the answer is somewhat subtle and in some ways uncertain: while, explicitly, mechanists have tended to distance themselves from analogies to machines and have stated that mechanisms are not to be identified with machines, the program has relied on examples that have a machine-like character, and this has shaped expectations and intuitions. We do not wish to argue that the cases we discuss violate the letter of definitions offered by recent "new mechanists," and we do recognize that the program was not meant to be restricted to machine-inspired cases. However, we also think that, beneath the surface, the discussion of mechanisms, both by proponents and critics, has been overly influenced by the machine image, and part of our goal is to weaken this nexus.

To be sure, we do think the notion of a mechanism is valuable, and we continue to rely on it. A loose but useful construal has it that a mechanism is any system whose behavior can be explained by appeal to its underlying composition, i.e., its components and the relations between them. This is a very general class of systems (and corresponding explanations). The category of machine-like systems is contained within it, as machines have specific sorts of internal structures. We will comment further on the relation between mechanisms and machines and on closely related questions – such as what a non-machine-like mechanism is – as we go along and especially in section 4. But we should emphasize that a detailed discussion of this issue is beyond the scope of this chapter (for that, see Levy 2014).

Some critics of mechanism may object that the inclusion of non-machine-like systems as mechanisms extends the notion of mechanism too far. The fear seems to be that such extensions threaten to treat all explanations as mechanistic. We offer three lines of response. First, the extension only includes explanations that appeal to underlying composition. There are many types of explanations, including teleological explanations, etiological explanations, and some mathematical and dynamical explanations, that do not appeal to constitution and so do not count as mechanistic in our view. Second, the notion of mechanism, as discussed explicitly by Bechtel and Richardson (1993/2010), is an evolving concept that scientists elaborate on as it serves their explanatory needs. Third, our inclusion of non-machine-like systems as mechanisms is motivated by the scientists who describe the types of systems on which we focus as mechanistic.

We begin in section 2 by characterizing the machine metaphor and discussing its value, methodologically speaking. Here and throughout we rely on examples to illustrate key points and to tie our discussion to concrete aspects of scientific practice, as is common in writing on mechanisms. Then, in section 3, we look at several ways in which research on mechanisms has come to diverge from the classic machine-inspired work, including cases where concentrations matter in relatively subtle ways and examples in which a mechanism's internal constitution or organization, or both, changes while the mechanism operates. Finally, in section 4, we discuss the relation between mechanisms and machines and provide an overview of the state of machine metaphors as we see them.

## The machine metaphor and accounts of mechanistic explanation

The machine metaphor has been part of philosophical thinking about natural systems, and organisms in particular, since at least Descartes. For Descartes and mechanists of his stripe, the appeal to machines was part of an overall philosophical method, which, famously, included a spare ontology of natural systems and a set of rules for inquiry.[3] Important to Descartes's explanatory vision was that complex phenomena, such as magnetism and animal responses to stimuli, were to be explained in terms of the way corpuscles, the foundational components of his physical ontology, are put together. The modern metaphor of a machine, which is our topic here, plays a substantial, albeit more circumscribed, role. First, while it maintains a focus on identifying components and how they are organized, it makes no commitments to specific metaphysical foundations other than that the components can be identified through empirical inquiry. Second, and relatedly, machines are seen as just one kind of natural system and not as a general model.

In modern thinking, describing a system as machine-like implies that it has a certain kind of underlying causal structure (Levy 2014; Nicholson 2012, 2013). If we consider a familiar machine such as a coffee maker or an internal combustion engine, we see a relatively well-defined object made up of a set of distinct parts, each with a specific shape, size, and location. As the machine operates, each part performs a certain operation or activity, and the parts are so organized that the activities of the different parts combine in space and time in specific ways to generate a phenomenon (say, producing coffee or generating directional motion). Speaking generally, in cases like this the system's boundaries and the set of parts and their manner of interaction are stable and do not change as the machine works. To understand such a system, one must primarily understand the manner in which labor is divided, so to speak, among its parts: what each part does and when and how each part's interaction with other parts (when it does interact) contributes to the machine's exhibition of its overall behavior.

There are biological phenomena that appear similar in their basic structure, calling for similar kinds of explanations. Consider the mechanism for protein synthesis situated in the cytoplasm of eukaryotic cells, whose discovery Darden and Craver (2002) have analyzed in detail. The central component of the mechanism is the ribosome, a structure to which individual tRNA molecules ferrying amino acids dock when they match the codon on the resident mRNA. Once the tRNAs dock, the amino acids they ferry can be successively added to the developing polypeptide chain. The activity of the ribosome is characterized in terms of both chemical bonds and physical movements of components. In the first frame of Figure 5.1, tRNAs are docked to the mRNA at both the E and P sites in the small unit of the ribosome. In the second frame, a new tRNA coupled to elongation factor EF-Tu-GTP binds to the mRNA at the open A site and the tRNA at the E site is released. The binding of the tRNA to the codon on the mRNA activates the GTPase within the ribosome that hydrolyzes GTP. The energy from hydrolysis releases EF-Tu-GDP and causes a conformation change in the rRNA that is part of the ribosome. This orients components so that the peptidyl transferase can transfer the peptide chain onto the tRNA at the A site. At this point, another molecule of GTP, bound to EF-G, binds, and as this GTP is hydrolyzed, it causes the mRNA and its attached tRNA at the P and A sites to move one location in the ribosome (to the E and P) sites.

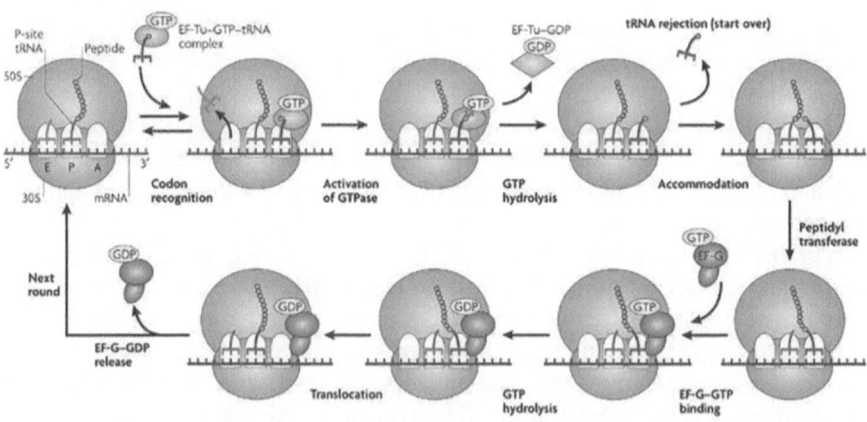

Figure 5.1 Steps in the synthesis of a peptide chain.

Source: From Steitz, 2008.

Note: This figure can be accessed in color via the eBook version of the book and eResources at www. routledge.com/9780815380788.

As can be seen, the physical movement of the tRNAs bound to mRNAs through the ribosome resembles the physical movement of components in human-made machines. In many ways, it is akin to a production line. Other mechanistic explanations developed in the 20th century, such as the ATP synthase that adds a phosphate to ADP, relying on a proton-motive force over the inner mitochondrial membrane, and the motor and flagellum that the bacterium *E coli* (as well as some other microorganisms) use to swim, exhibit a broadly similar structure.[4]

Bechtel and Richardson (1993/2010) characterized mechanism discovery as starting with the delineation of a phenomenon and associating it with a system they took as the locus of control for the phenomenon. Researchers then made the heuristic assumption that the system consisted of parts that performed operations that could together produce the phenomenon and used various strategies to decompose the system physically or the phenomenon functionally. In many cases, researchers started with tools for either structural or functional decomposition, but ultimately they sought to localize operations in specific parts. As developed by Bechtel and Abrahamsen (2005), after decomposing a mechanism, researchers need to recompose it (sometimes by mental rehearsal, sometimes in computer simulations) to show that when appropriately organized the parts and operations can generate the phenomenon. Darden (2002; developed further in Craver and Darden 2013) adds additional strategies, including forward/backward chaining and assembly of modules. The first of these supplements strategies for decomposition, leading investigators, after they have identified one or a few activities in a mechanism, to look for clues as to what could precede and what could follow those activities. The assembly-of-modules strategy supplements recomposition by encouraging researchers to operate much like designers, putting together modules with known functions in novel arrangements to account for new phenomena.

The new mechanists developed these accounts by drawing on specific examples from cell and molecular biology, such as protein synthesis, long-term potentiation, glycolysis, and oxidative metabolism. In the accounts of the biologists, each of these phenomena results from the operation of a distinct mechanism that is localized in the cell (or one of its organelles). Research strategies for decomposing the mechanism assumed that it was an enduring entity that could be broken down into components. The components themselves were also taken to be enduring and organized so as to feed the outputs of the activities of one component into another until the whole activity was performed. This is readily reflected in the discovery strategies described earlier: modular assembly assumes that component mechanisms can be combined so that the activity of the whole is a straightforward (in some cases additive) product of the activity of the components. Forward/backward chaining assumes that there are distinct components, ordered sequentially, such that the outputs of some such component are fed into the known component and it, in turn, provides its product to one downstream of it.

A further aspect of research under the machine analogy concerns the favored means of representation. Much as engineers explore and develop their designs in diagrams, biologists often use mechanism diagrams to reason about and guide investigation (Abrahamsen et al. 2018). In these diagrams, parts, operations, and organization are represented in two (sometimes three) dimensions. The parts are the easiest to incorporate into diagrams, as they can be represented with shapes, which might be iconic, or words. Consider Figure 5.2, which represents the process of long-term potentiation wherein the response of AMPA receptors at post-synaptic cells is enhanced as a result of influxes of calcium due to activation of NMDA receptors. The receptors are shown with iconic symbols, while the molecular intermediates are indicated by their names. Operations are often more difficult to represent, but typically arrows are used in some way. In fact, arrows often serve multiple different functions. The arrows passing through the AMPA and NMDA receptors indicate the passage of Na+ and Ca2+ into the cell, whereas the arrow from CaMKII to the AMPA receptors indicates phosphorylation of the AMPA receptor (one of two ways of enhancing the response of AMPA receptors). The arrows from Ca2+ to Ca2+/calmodulin and then to CaMKII indicate chemical reactions in which Ca2+ binds to calmodulin, which phosphorylates a site on CaMKII that terminates its autoinhibitory activity and renders it an active kinase able to phosphorylate AMPA receptors. All of these components are shown in the

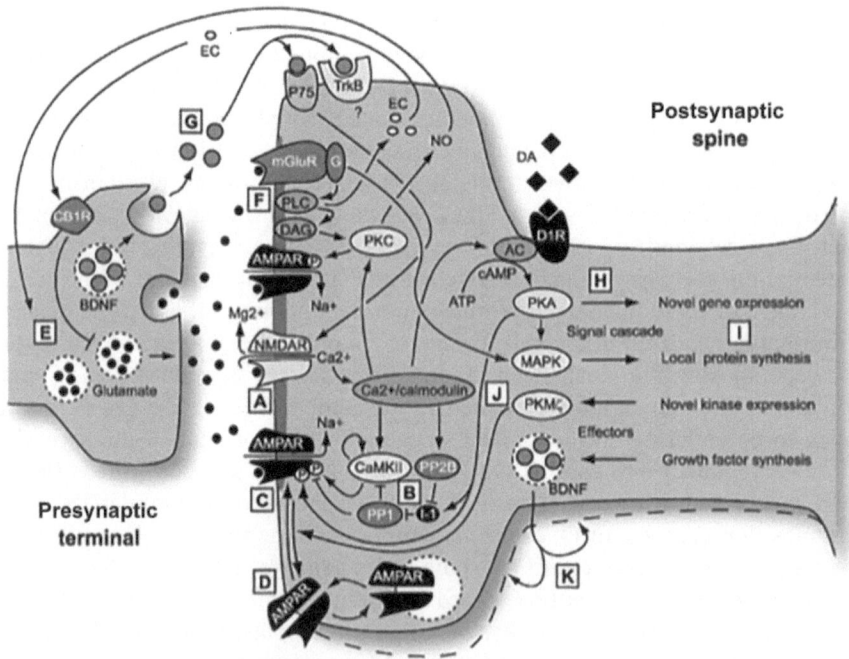

*Figure 5.2* The core mechanism of long-term potentiation.

Source: Figure reprinted from Bliss, T. V. P., & Cooke, S. F. (2011). Long-term potentiation and long-term depression: a clinical perspective. Clinics, 66, 3–17 and distributed under Creative Commons License (CC BY-SA 3.0).

postsynaptic terminal, with the arrows showing how they are organized functionally with respect to each other.

Diagrams do more than represent hypotheses about mechanisms. Scientists often use them in the process of discovering and reasoning about mechanisms. One sign of the role diagrams play in discovery is the not infrequent occurrence of question marks in mechanism diagrams. These are often used to signify recognized gaps or hypotheses that have not yet been sufficiently supported experimentally (Bechtel et al. 2018). In Figure 5.2, a question mark is shown by the TrkB receptor, indicating it was assumed to be involved, but it was not known how. The diagram provides a way of representing hypotheses that require further investigation.

Diagrams also support reasoning about how a mechanism produces a phenomenon. Static diagrams, unlike animations, do not explicitly show how the mechanism works. A common way to reason from a static diagram employs what Hegarty (1992) characterizes as mental animation: the viewer of the diagram mentally rehearses the activities that are shown. Figure 5.1 illustrates a different strategy – showing sequential steps in different frames. In that figure, the arrows between the frames show the sequence in which the steps occur, ultimately returning to the initial state. In some cases, such as when multiple activities are occurring at the same time or at different time scales, this is too demanding and researchers turn instead to computational simulation.

Figures 5.1 and 5.2 reinforce the features we have attributed to the machine conception of a mechanism. In each, both the mechanism and its parts are presented as localized and enduring entities engaged in specific activities conveyed by the arrows. Moreover, the arrangement of parts is stable, and this is crucial to the activity of the mechanism.

## Mechanistic accounts that depart from the machine metaphor

In this section, we discuss two dimensions along which research on mechanisms in contemporary biology has moved beyond the machine metaphor: first, in targeting systems in which the working parts are distributed (rather than localized) and where the activities performed depend on concentrations and related "bulk" properties; and second, in studying systems in which the menu of parts or their organization, or both, changes over time, requiring dynamical analysis of constitutional change. In both cases, as we shall see, a key issue is the need to track dynamical and quantitative properties in ways that were less important or altogether absent in earlier, machine-inspired research.[5]

As with the machine conception of mechanisms, diagrams are often invoked to represent and to reason about less machine-like mechanisms. However, to handle the more quantitative and dynamical features of these mechanisms, different diagramming techniques are employed. The diagrams we present from the research discussed in this section illustrate some of the strategies for adapting diagrams to represent dynamical mechanisms.

### Mechanisms whose behavior depends on concentrations of working parts

The components in mechanistic explanations are often treated as if there were one copy of each. This suggestion is also conveyed by mechanism diagrams, such as Figure 5.2 and several below. But this is typically incorrect. While an organism may have one heart or two kidneys, a cell typically contains multiple – oftentimes a great many – copies of each molecular species.[6] More importantly, in some cases the ability of these components to perform their activities depends on their prevalence – either their absolute numbers or, more frequently, their concentration. Broadly speaking, we can distinguish two aspects of tackling distributed components. First, the simple fact that such components are inherently nonlocalized requires the adoption of different methods of representation and analysis. Second, in cases of distributed working parts, often the properties of greatest importance for explaining a particular phenomenon are quantitative, "bulk" properties – especially measures of relative abundance in space and time, such as densities, concentrations, copy numbers, and the like – rather than structural-qualitative.

To be sure, distributed components figure explicitly in many machines and machine-like mechanisms. The liquid that flows through the coffee maker introduced earlier consists of a multitude of molecules. Likewise, in biochemical reactions a distributed collection of molecules is transformed. But in these cases, the role of distributed components is relatively straightforward, and the precise details, in particular precise quantitative aspects, are not focal. In these cases, the distributed substances are operated on, but they are not *performing* operations. In other cases, the distributed components are working parts. In a steam engine, the steam, consisting of distributed molecules, pushes the fans of the turbine. The quantity of molecules determines how fast the turbine moves. Similarly, in the mechanism of oxidative phosphorylation, the number of protons crossing the membrane at the ATPase determines the number of rotations of the windmill-like rotor that figures in the addition of phosphates to ADP. Still, in these cases the effects of the whole are simply the sum of the effects of the individual particles. One needs a measurement of the total number of molecules to understand the magnitude of the effects, but one does not require exact quantitative modelling to develop sufficient qualitative explanations of how the mechanism works. The mechanism still operates in a machine-like manner, in that the multitudes of distributed components are treated as one "big" component, with properties (such as sheer magnitude) that fit comfortably within machine-like thinking.

By contrast, the machine perspective breaks down when the operation cannot be understood as simply a sum of the individual effects. We consider cases of this sort wherein even subtle differences in the distribution and relative magnitude of a component make an essential difference to what happens in the mechanism; without attending to these aspects one cannot make sense of the phenomenon. The example we consider involves genetic switches, dynamical molecular mechanisms with multiple (typically two) stable states, in which it is possible for the mechanism to be in either one of them. Under certain conditions the switch will move from one state to the other in a largely discontinuous manner. For instance, the system may move from a state in which it synthesizes little of a given mRNA to one in which a relatively high level is synthesized. Such switches play important roles in developmental contexts, where setting the switch to a specific state constitutes a "decision" that sends a particular tissue on a distinct developmental path.

Using genetic switches as an example, Nathan (2014) drew philosophers' attention to cases in which the causal efficacy of concentrations is not just the sum of the efficacies of individual components. No single molecule is responsible for the state of the switch. Multiple molecules must act in the same direction during the relative time window. It is the concentration that determines which state the switch is in. Switching, as opposed to a continually varying response, is due to feedback loops (positive and negative and often both), which generate thresholds. Negative feedback may tend to keep concentrations below the threshold, but if they happen to cross the threshold, positive feedback may result in the system "snowballing" into a different state.[7]

Molecular switches are important – and best studied – in microorganisms in which they enable the organism to undergo rapid metabolic changes in gene expression. Consider, for instance, the lactose (lac) operon in *E coli*. This is one of the best studied systems in biology and one of the most discussed examples in philosophical contexts. These discussions, however, have focused on some of the basic features of the operon that can be treated as the sum of individual molecular processes, not its threshold behavior.

We review first those features of the operon that can be understood as a sum of individual molecular activities before turning in subsequent paragraphs to the threshold behavior that has been explored only in recent years (partly due to the availability of novel methods, on which we comment later). Originally discovered by Jacob and Monod (1961), the lac operon is a stretch of bacterial DNA that encodes enzymes necessary for the metabolism of galactose, a sugar that the bacterium consumes when, and only when, lactose is absent. When the need to metabolize galactose arises, bacteria switch to producing the relevant enzymatic machinery. Structurally, the system's basic elements are depicted in Figure 5.3.[8] When glucose is abundant, the lac genes are inhibited. This is achieved through a repressor, *lacI* (number 2 in the image), blocking the relevant genetic operator (number 4). With the repressor in place, RNA polymerase cannot transcribe the lac genes. When galactose is present, it binds to the repressor, thus neutralizing it and allowing the transcription of lac genes to commence. To indicate that there is not just one molecule of galactose, multiple copies of the cross-shaped icon are shown, but the crucial reaction is presented as involving one repressor operating at one locus. The removal of the repressor leads to the synthesis of new enzymes relevant to galactose metabolism, such as β-galactosidase (encoded by *lacZ*) and β-galactoside-transacetylase. *LacY* (number 7) is also on the operon; it encodes a transmembrane pump, β-galactoside permease, that pumps more lactose molecules into the cell. The increased intracellular concentration serves to increase the rate at which lac genes get transcribed. When galactose is depleted, the same process works in the reverse direction, moving into an "off" state.

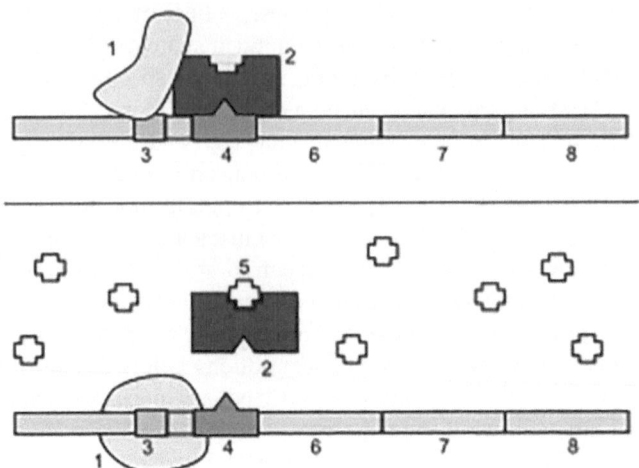

*Figure 5.3* Part of the lac operon that limits transcription of the lac genes (6, 7, 8) to situa-
tions in which galactose (5) is present and able to bind to the lac repressor (2),
removing it from the operator.

Source: Figure by T. A. Raju and distributed under Creative Commons License CC BY-SA 3.0.

Note: This figure can be accessed in color via the eBook version of the book and eResources at www.
routledge.com/9780815380788.

Most discussions of the lac operon in philosophy of biology focus, as we did
in the previous paragraph, on individual molecules that bind or detach from
the switch. But whether the switch remains in one state or changes to the other
depends on the relative concentrations of different molecules. Moreover, it is not
merely the ability of an inducing or inhibiting molecule to bind to DNA – and to
initiate transcription – that explains switching. Indeed, inducers and inhibitors
are bound to DNA, to some extent, at all times. Rather, it is the relative rate of
binding that determines the relative concentrations of inducers and inhibitors, and
it is these that constitute the state of the switch. What requires explanation is the
switch-like behavior in which the system quickly transitions between states. One
way to appreciate this behavior is to consider intermediate concentrations. In such
conditions, one doesn't get intermediate levels of transcription, as one would if
what mattered were just the gradually increasing concentration. Rather, each cell
has either very few or many lac proteins. And each cell transitions abruptly. If one
examines a population of genetically identical cells, one sees that some cells have
switched, producing high levels of lac proteins, while others haven't.[9]

The account of the lac operon provided so far cannot account for this switching
behavior. What accounts for the switching behavior is that transcription does not
increase smoothly but exhibits sporadic "bursts" of transcription from the operon.
If these bursts are large enough, they move the cell into a different state. Bursting

is inherently a quantitative phenomenon, as a burst is defined as an increased transcription of a gene for a short period of time. While transcriptional bursting has been known for decades and is a prevalent phenomenon (it varies according to the organism, the gene in question, and other factors; see Lenstra et al. 2016), its mechanistic underpinnings and how they generate discontinuous switching behavior are only beginning to come into view. We discuss some recent work on bursting in the lac operon, highlighting ways in which this sort of phenomenon requires different investigative strategies relative to work on machine-like systems.

Choi et al. (2008) demonstrated that burst size is crucial to whether a cell switches: only large bursts (<200 copies transcribed) trigger a positive feedback loop that maintains a high rate of transcription that constitutes being in the switched on state. Small bursts of up to 10 transcripts are much more frequent but do not result in positive feedback or switching. Choi et al. then investigated, both theoretically and experimentally, what produces the large bursts. In their theoretical analysis, the researchers considered two classes of models. In one, large bursts are due to multiple events, each with a relatively large probability. Many events are required to generate a quantity of transcripts that exceeds the switching threshold. In the other, the large burst arises from a single event that is sufficient to raise the concentration above threshold. Both models predict the same switching rate. To decide between them, Choi et al. appeal to robustness, i.e., the idea that switching is not affected by changes in cellular componentry (e.g., ribosome numbers and the efficiency of RNA polymerase) other than those specific to lac transcription. On their theoretical analysis, only the single-event model displays robustness. Accordingly, the researchers concluded that a single, rare molecular event is responsible for bursting and hence switching.[10] It is noteworthy that even though the researchers conclude that a single event induces switching, to show this they had to analyze the quantitative character of the system.

Choi et al. went on to investigate which molecular event was responsible. First, they ruled out an older theoretical proposal to the effect that it involved the permease (encoded by *lacY* gene), i.e., the membrane-bound pump that lets more galactose into the cell. To do this, they attached the fluorescent protein YFP to the permease (Xie et al. 2008 provide a review of such methods) and showed that switching occurred with only a fairly large number of permease molecules (on the order of 300). There was no evidence of bursts of pump production that would lead to such large copy numbers of the permease.

As an alternative, the researchers investigated LacI, a tetramer that binds to two sites on the DNA so as to create a loop. They hypothesized that dislocation from one site produces the short burst. Since dislocations from both sites require several minutes for the repressor to rebind to both sites, they reasoned that it would result in longer bursts. As shown in Figure 5.4A, when the concentration of the inducer is high, the repressor is effectively removed from the operator at both sites, allowing transcription of *lacI*. When the concentration is lower, the repressor is generally removed at only one site, resulting in a small burst. On some occasions, though, even with a low concentration of inducer, the repressor is removed

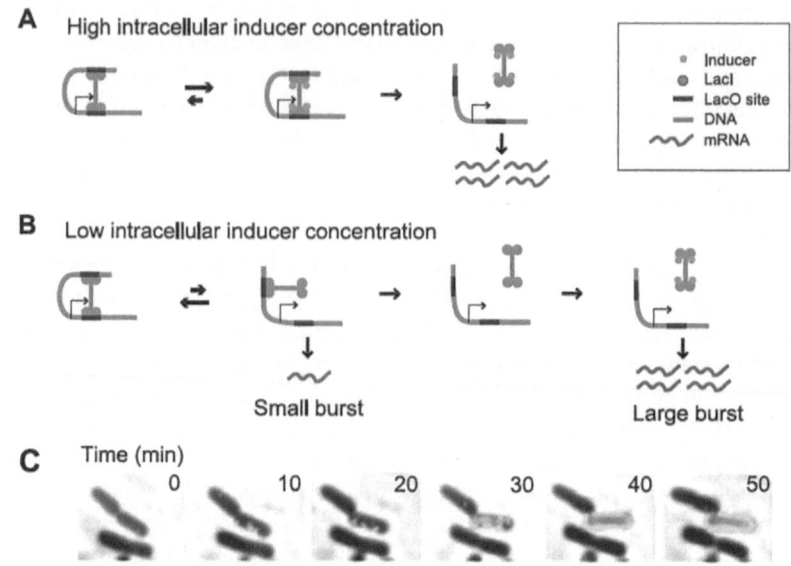

*Figure 5.4* Generation of small and large bursts in regulating the lac operon.

Source: From (Choi et al. 2008).

Note: This figure can be accessed in color via the eBook version of the book and eResources at www. routledge.com/9780815380788.

at both sites. It takes much longer to reattach at both sites, and this results in a large burst (Figure 5.4B). Choi et al. summarize their model with reference to Figure 5.4 as follows:

> Under low or intermediate [inducer] concentrations [. . .] the repressor can stochastically dissociate from one operator independently of the inducer, as shown in Fig. 4B. When the repressor partially dissociates from one operator, a small protein burst from a single copy of mRNA is generated before the repressor rapidly rebinds to the vacant operator. When the repressor completely dissociates from both operators, multiple mRNAs are transcribed, which leads to a large protein burst that surpasses the LacY threshold, initiates positive feedback, and maintains a switch in phenotype.
>
> (2008, 445)

The researchers also offered several sources of empirical support for their hypothesis. One was a relatively straightforward approach to investigating mechanisms: remove a critical component and show that this impairs the phenomenon. In this case, the researchers observed that mutant strains that do not exhibit the DNA loop also do not show bistability. Instead, the frequency of large bursts is high and switching probability is proportional to inducer concentration. The other evidence they adduced, however, involves time series measurements of quantities. For example, the researchers replaced the *lacY* gene with a gene encoding tsr, a membrane-bound protein having nothing to do with lac metabolism, bound to YFP. First, this allowed them to eliminate positive feedback. Second, the fused protein served as a reporter molecule, allowing an estimation of burst size by counting the number of tsr-YFP molecules present at the membrane over time. Frequent small bursts and rare large bursts were observed.

There are several points we wish to highlight by considering this example. First, the phenomena at issue – switching and especially bursting – are essentially quantitative. While structural aspects of the process are crucially important – the mechanism of association and dissociation at one or both operators is key – what they account for are transcription bursts that potentially send the system above threshold. It is the bursts that constitute the switch. Second, and most importantly for our present purposes, at both the theoretical and the empirical level, studies of this sort use tools that allow one to resolve quantitative questions in both time and space. At the theoretical level, as we have described, researchers relied on robustness as a criterion for model selection. At the empirical level, tools for measuring copy numbers for specific molecules and for obtaining time series data were essential. These tools for discovery and confirmation are essentially tied to the fact that the mechanism at issue relies on bulk properties – concentrations, copy numbers, etc. – backgrounding structural matters. Finally, the discontinuity between high and low concentrations of the inducer made it impossible for the researchers to represent the mechanism in one mechanism diagram. Rather, the researchers needed to present the mechanism in two different panels, each showing its behavior in one of the conditions. We revisit these points in section 4. But before that we will look at another respect in which research into mechanisms diverges from the expectations generated by machine analogics.

### Mechanisms as dynamically changing

Relying on the machine metaphor, biological mechanisms are viewed as like the machines of a factory, sitting and waiting to be "turned on" – i.e., for their start-up conditions to materialize. They then carry out their sequential activities until they reach their termination conditions (a widget being produced, etc.), at which point their output is passed to another mechanism that carries out a different activity (wrapping the widget, say). Many biological mechanisms are not fixed entities but dynamically changing systems. In this section, we briefly present three examples of mechanisms undergoing dynamical change and then turn to the consequences for thinking about mechanism discovery and associated methodologies.

We start with an example of what might seem a rather modest break from the machine conception of mechanism and then move to more radical examples. On the machine conception, components of the machine exist in a specific organization, passing intermediates between them. But in some cases, the way the parts are organized with respect to each other is radically altered during the course of the mechanism's operation, and this affects further operations the mechanism performs. This is exemplified in the circadian clock – a mechanism that, like a human-made clock, cycles through states that correspond to different periods of the day. In cyanobacteria, time of day is indicated by the phosphorylation state of KaiC molecules.[11] Shown as a green donut in Figure 5.5, KaiC is a hexamer, and each monomer has two sites that can be phosphorylated. KaiC is both an autophosphatase and an autokinase – i.e., it can catalyze both of these reactions itself. Which is favored depends on how two other proteins align themselves with KaiC. It is the particular organization of components at a given time that determines whether KaiC procures phosphate groups from ATP or releases them.

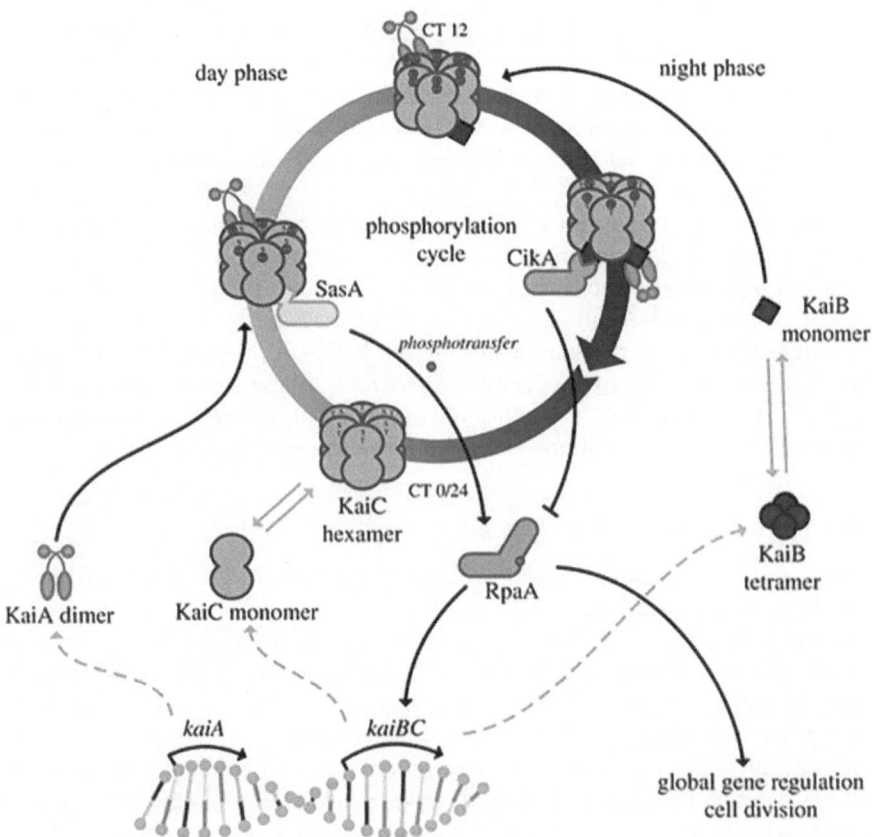

*Figure 5.5* The core circadian clock mechanism in cyanobacteria.

Source: Figure reprinted from Schmelling, N. M., & Axmann, I. M. (2018). Computational modelling unravels the precise clockwork of cyanobacteria. Interface Focus, 8, 20180038 and distributed under Creative Commons License (CC BY-SA 4.0). This figure can be accessed in color via the eBook version of the book and eResources at www.routledge.com/9780815380788.

To convey such reorganization, Figure 5.5 shows the mechanism's configuration at four different stages, which correspond to the clock registering different times of day (indicated by the shading of the overall circle). At the stage shown on the left (corresponding to the clock signifying morning), the A-loops on the CII domain of KaiC hexamers are extended, affording sites at which KaiA can bind. When Kai A binds, it favors phosphorylation of KaiC. But as a result of this phosphorylation, the conformation of KaiC changes, enabling KaiB to bind to the CI domain. At the same time, the A-loops on the CII domain are retracted. As a result of these two operations, KaiA becomes bound to KaiB and unavailable to promote phosphorylation. The key thing to note in the diagram is that KaiC has different relations to KaiA and KaiB at each of the four steps and that this is what determines whether phosphates are taken up or released. This brings out the point that it is the way the mechanism is organized at each period of the day that determines how it operates. The organization of the mechanism is transitory, and to explain the mechanism's functioning, researchers need to consider how it is organized at each time period.

Organizational changes as the mechanism operates may be seen (we are unsure) as a relatively modest challenge to the machine conception. The idea that the parts of a mechanism change over time represents a more radical change. Recognizing that parts change is difficult when one pursues the traditional strategy of first identifying a mechanism and then decomposing it. But in recent years some researchers have approached mechanisms in a different way. They start with networks of related components, represented as nodes connected by edges, in a network representation. In many studies, researchers use evidence that proteins can interact as the basis for determining when they should be linked by edges.[12] They then invoke algorithms to identify clusters of nodes (often called "modules") that are interconnected much more often than is typical in the network. Insofar as the nodes are highly interconnected, researchers infer that the entities involved might work as mechanisms. Some initial network representations ignored the timing of the interactions, but since many researchers were interested in how cells adapt to specific challenges, investigators began to employ time series data. Let us first describe one such study in more detail and then comment on how the methods involved in it are novel relative to studies that employ machine-style strategies.

In their study of the cell cycle, de Lichtenberg et al. (2005) annotated a protein-protein interaction network with data about timing of transcription, generating the network representation shown in Figure 5.6. The network contains 300 proteins, of which 184 are expressed only during specific phases of the cell cycle. Color is used to indicate time of peak expression (white is used for proteins that are expressed constitutively.) To link color to the phase of the cell cycle, the 412 proteins for which the researchers could identify no interactions are shown around the outside of the circle in the color assigned to the phase of their peak expression in the sequence of phases of the cell cycle. Thus, to determine the phase of expression of proteins that are parts of connected networks, one needs to match the color to the color of the nodes around the outside. Many of the modules shown

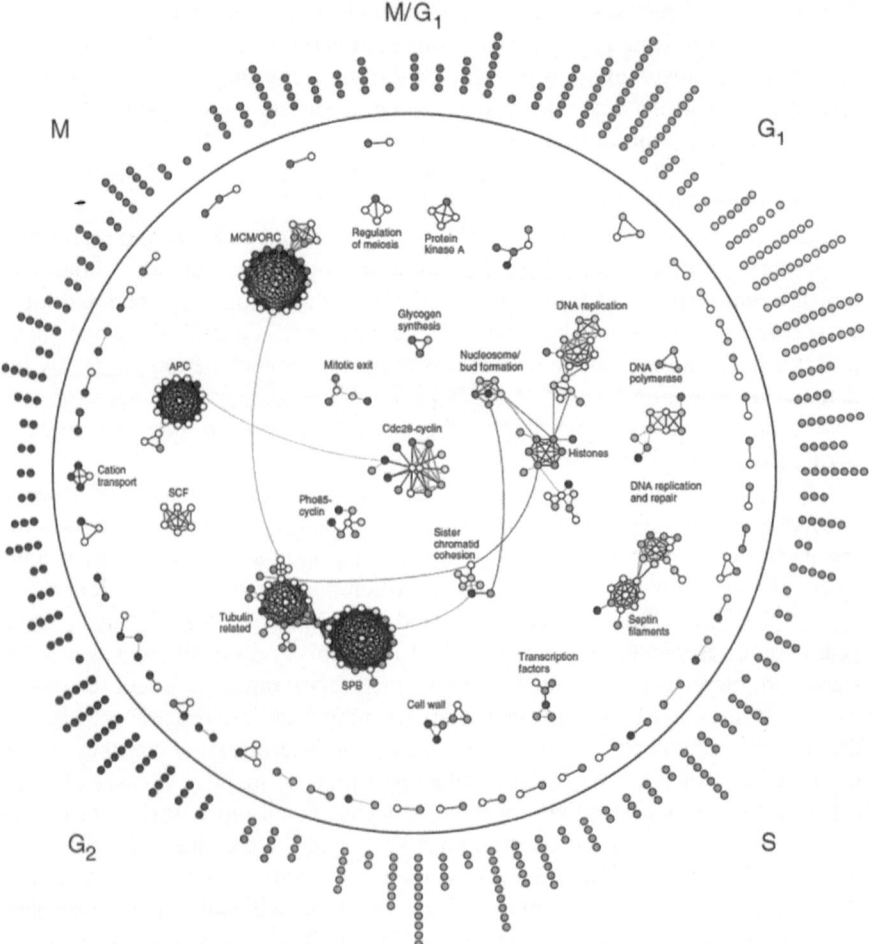

*Figure 5.6* Network analysis of genes expressed at different phases of the cell cycle.

Source: From de Lichtenberg et al. (2005).

Note: This figure can be accessed in color via the eBook version of the book and eResources at www. routledge.com/9780815380788.

in the center of the network diagram could be associated with known functions in the cell cycle and so correspond to mechanisms. Numerous components of these mechanisms were already known from more traditional mechanistic research, but the network approach did allow de Lichtenberg et al. to locate 30 proteins not previously associated with the cell cycle within modules/mechanisms in the cell cycle and to link several other proteins whose role in the cell cycle was poorly characterized to specific modules/mechanisms.

It is worth emphasizing how Figure 5.6 differs from typical mechanism diagrams. For one thing, it does not specify specific operations for the various entities identified – edges simply indicate that the proteins are capable of interacting. Second, it shows change in the mechanism through the stages of the cell cycle by the use of color. The utility of this approach is illustrated when de Lichtenberg et al. focus in on three specific modules/mechanisms in Figure 5.7.

These individual modules illustrate a common pattern: some of the proteins in a module are constitutively expressed, but one or more are just expressed at the time at which the activity associated with the mechanism is required, resulting in what the authors refer to as "just in time assembly." The left side of Figure 5.7 shows the assembly of the pre-replication module, which is built around the six Orc proteins that are bound to the origins of replication throughout the cycle. The complex shown on the upper left, including the Mcm proteins, is recruited to the proteins at the origin of replication by Cdc6p, which, like several of the proteins in the Mcm complex, is synthesized only during the G$_1$ phase. Final assembly then depends on Cdc45p, which is expressed early in the S phase. The right side of Figure 5.7 reflects a different pattern in which the cyclin-dependent kinase Cdc28p forms complexes at different phases of the cell cycle with different cyclin partners and inhibitors, capturing the known dynamical behavior of this mechanism in different activities of phosphorylation and ubiquitination at different phases of the cycle. The question mark accompanying Ypl014p reflects the researcher's tentative assignment of this previously uncharacterized protein to this known mechanism.

All of the modules/mechanisms identified by de Lichtenberg et al. exhibit different parts at different times. They are not, as is typical in machines, static entities with a fixed set of parts awaiting their start-up conditions to spring into action. Rather, their constitution changes over time as a function of the activity of the mechanism itself. When the parts of a mechanism change over time, identifying

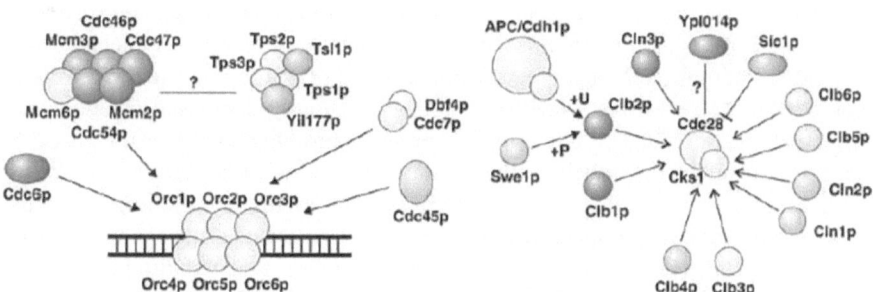

*Figure 5.7* Assembly of the pre-replication module (left) and of the interaction of Cdc28 with different partners at different phases of the cell cycle.

Source: De Lichtenberg et al. (2005).

Note: This figure can be accessed in color via the eBook version of the book and eResources at www.routledge.com/9780815380788.

the components and what they do does not suffice for explaining a phenomenon. One must determine the pattern according to which the components come into or go out of existence and, ideally, understand the process through which they come into or go out of existence. Internal-organizational change over time, which did not figure prominently in machine-inspired studies, now plays an increasingly important role. Moreover, this role was not discovered by the research strategies traditionally deployed by mechanists. Rather, the time-dependent character of the parts of these mechanisms was discovered by integrating the tools of network analysis, which enabled discovery of modules within interconnected components, with time series data about when proteins are expressed.

The de Lichtenberg example brings out a further dynamical aspect. In many cases the mechanism itself comes into existence only as needed and cannot be seen as a preexisting substructure of living systems. In the de Lichtenberg example, these components were generated in a regular manner through the cell cycle, but in other cases the very existence of the mechanism is found only in specific conditions that the cell might encounter only occasionally, if ever. To study, indeed to identify, such mechanisms requires strategies of a new sort, such as *differential network biology*, characterized by Ideker and Krogan (2012). In implementing the strategy, Bandyopadhyay et al. (2010) created a gene network in which the relations between 418 genes were based on whether two genes, when deleted together, had an effect on colony growth that was greater or less than expected from the effects when knocked out individually (pairs that exhibit such interactive effects are known as "synthetic lethals" since, in the simplest case, neither mutation alone is lethal, but the combination is). Bandyopadhyay et al. generated synthetic lethal networks for yeast grown under normal conditions and in a condition in which methyl methanesulfonate (MMS), a DNA-alkylating agent, was added to the medium. The two networks differed substantially, and to identify the differences, the researchers subtracted one network from the other to generate a so-called differential network. They found many gene pairs that could be detected only as synthetic lethals in the network as a result of the subtraction. In particular, the differential network revealed many interactions between DNA damage-response genes, suggesting that the proteins from these genes work together in mechanisms in the MMS condition. When the network was mapped onto one showing protein interactions, the researchers identified that differentially decreased or enhanced synthetic lethal interactions occur mainly between protein complexes. (Figure 5.8 shows in red reduced synthetic lethal connections between the HIR-C transcriptional corepressor complex and the INO80/C chromatin remodelling complex but in green increased connections between HIR-C and two other modules.) The researchers propose that different protein complexes are recruited into functioning mechanisms in the face of DNA damage produced by MMS.

As with the previous example, Figure 5.8 differs from a traditional mechanism diagram, in that the edges do not represent specific operations affecting other components but protein-protein interactions (black edges) and synthetic lethality (colored edges). Color in this case is used to indicate whether under the conditions inducing DNA damage synthetic lethality is increased (green) or decreased (red).

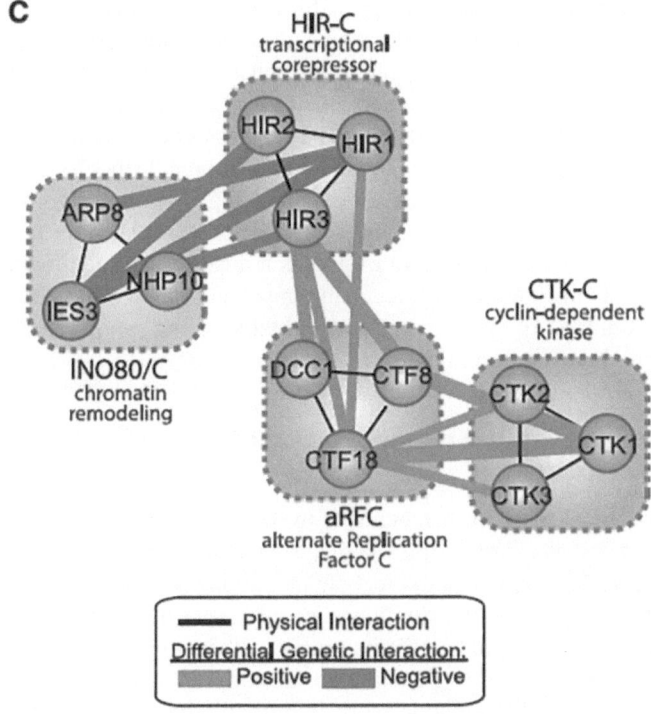

*Figure 5.8* Differential genetic interactions between different modules when yeast is grown in conditions generating DNA damage than when grown in normal media.

Source: From Bandyopadhyay et al. (2010).

Note: This figure can be accessed in color via the eBook version of the book and eResources at www. routledge.com/9780815380788.

In mechanistic terms, the cluster of nodes linked with green edges is viewed as interacting *as a mechanism* under DNA damage conditions, and those with red edges are viewed as not operating as a mechanism under those conditions. The diagram allows researchers to represent transient mechanisms that arise as the cell faces different challenges.

Thus, we see that cases in which mechanisms or their components come into or go out of existence require different tools and investigative strategies, both at the experimental and discovery level and at the level of representing results. One must find ways of tracking the system's structure over time to identify which parts are active when. Differential network biology, discussed previously, is one such method. It allows researchers to infer which sets of components – and by implication which potential mechanisms – are present in the cell at different points in time or under different conditions, or both. The graphical representations associated with such systems also take a different form relative to machine-like mechanisms:

they must represent the system's differing structure over time and show how either parts or organizational features change as the mechanism operates.

The assumed static nature of mechanisms is one source of criticism of the mechanistic framework leveled by theorists such as Nicholson. They argue that to account for the ability of biological organisms to maintain themselves and reproduce, the components of organisms must be seen as dynamic and shaped by their role in the whole organism. We cannot develop this topic further here, but if biologists' understanding of mechanism allows for changes in organization, then phenomena such as self-maintenance and reproduction can plausibly be accounted for mechanistically. As the last two examples indicate, such modifications of the concept of mechanism are in fact occurring in present-day biology. Biologists are increasingly thinking of mechanisms as metastable, being synthesized and organized when conditions require them.

## Conclusion: mechanistic research beyond the machine metaphor

Most philosophical work on mechanisms to date has explored a set of scientific strategies centered on a particular class of systems; namely, machine-like systems. We view this work as highly valuable, both in drawing attention to and explicating the relevant strategies and in providing a counterpoint to more traditional views of scientific explanation, and scientific practice more broadly, that center around laws, logical structure, and explanation as subsumption. However, we also think that the purview of machine-style work is limited, and in this chapter we have sought to move beyond it, noting cases where core features of machine-like systems are absent but researchers still invoke mechanisms in their explanations. We characterize some of the strategies used by scientists who explore such systems.

As we noted early on, we are well aware that mechanists have explicitly cautioned against identifying mechanisms with machines. As we see it, one should distinguish a broad and relatively minimal sense of mechanism, i.e., a complex system with multiple parts that come together to generate a phenomenon (cf. Glennan 2017). A mechanistic explanation, accordingly, is an explanation that reveals the underlying parts and how they come together to generate the phenomenon. But while this is the "official" sense of mechanism that most mechanists have assumed, many of the examples they have brought forth, and many of the specific claims about scientific practice they have made, in fact apply to machine-like cases. We do not view this as a serious criticism or as pointing to a deep problem. But we do think it calls for an expansion of focus in discussions of mechanistic explanations.

In arguing for such an alteration of focus, we have highlighted two out of a multitude of features that machine-like systems exhibit: localized parts whose structural aspects are explanatorily key and parts and organizational features of the mechanism that remain stable throughout the mechanism's operation. There are other ways in which mechanistic science departs from the machine metaphor, but to keep the discussion focused we have decided to dwell on these.

The first aspect we have highlighted concerns the character of components: standard machines tend to have parts with a stable location and timing of operation and in which the structural aspects are those that do the causal-explanatory work. By contrast, we have looked at cases in which parts are distributed in space or time, or both, and where "bulk" properties, like concentrations and rates, are explanatorily central. Genetic switching is an illustration of such a phenomenon: switching itself is due to the balance of concentrations of the relevant chemical species – an inhibitor, an inducer, and a molecular pump, in the case of the lac operon we discussed. Moreover, to understand intermediate regimes, where a population of bacteria exhibits a polymorphism with respect to switching, one must attend to transcriptional bursting, which by its very nature is a rate-related phenomenon. As we have seen, studying such phenomena involves different discovery and confirmation strategies, and the resultant explanations are distinctive. Measurements take on a much more rigorous quantitative character, methods for tracking single molecules are necessary, and criteria for selecting among models or of fixing model parameters (such as robustness) are required.

A second aspect concerns the stability of the system's internal composition. In machine-like systems, parts and their organization tend to remain in place while the mechanism operates. But there are many important exceptions to this in biology. In many cases, parts are formed and destroyed as the mechanism works. Likewise, in many cases the system's organization is modified as part of the mechanism's operation. We saw several examples in this vein involving central cellular functions, such as the cell cycle and the circadian clock. In studies of these phenomena, we saw that researchers adapt to the need to track how the system's internal composition changes over time via adding time series data to protein-protein interaction data that enable researchers to analyze how the mechanism comes together in time and via comparative network studies that allow them to discern how the system's composition changes when it is placed under different conditions.

One way to appreciate the differences between these mechanisms and machine-like mechanisms is to consider the differences in the way researchers diagram them. Standard mechanism diagrams emphasize the enduring components and the specific operations in which they are involved. These correspond in many ways to the schematics of traditional machines. But such representations are insufficient to represent and reason about the mechanisms we have been considering. Accordingly, we noted some specific ways in which biologists have extended their repertoire in diagramming these mechanisms. In Figure 5.3, the importance of multiple molecules of galactose is shown by including multiple copies of the icon representing it when it is present. To contrast how different inducer concentrations result in different bursting patterns, the authors of Figure 5.4 resorted to separate panels. As network diagrams, Figures 5.6–5.8 are less informative (relative to standard mechanism representations) with respect to showing operations. Edges show only that proteins can interact or genes that are synthetically lethal. To indicate how different components are synthesized at different times or interact differently in different conditions, the researchers use color. The fact that researchers have deployed novel visual representational techniques provides further illustration (no pun intended) of the ways in which these mechanisms depart from the machine model.

As noted, we do not take ourselves to be offering a refutation or even much of a criticism of existing work on mechanisms, but rather advocating an expansion in focus. To be sure, there are other aspects of machine-like systems that we have not looked at, and *a fortiori* there are potential deviations from machine-likeness that we have not discussed. The attempt to expand our thinking about mechanisms beyond the machine metaphor is in its early stages. Our goal has been to bring attention to some example explanations that do not fit the machine model and to the rich array of methods researchers have developed to investigate and represent them.

## Notes

1 This is true even as the conception of what a machine is has expanded in the 20th and 21st centuries with inventions such as the digital computer.

2 From now on, when we talk of machines we have in mind the machines of the industrial age that exemplify these features. Like biological evolution, machine designers are not restricted to this traditional conception of a machine and have developed many machines that violate these constraints.

3 For philosophers, a central feature of Descartes's account is his dualist metaphysics. Regardless of whether one takes Descartes's commitment to dualism as foundational or as simply a result of being unable to imagine machines that could produce particular phenomena – as we are inclined to view the matter – his views on the nature of the soul are secondary to his mechanistic project.

4 It is notable, with regard to many of these examples, that the extremely machine-like character of these mechanisms was not anticipated as researchers began their work. Only as they began to take components such as the ATPase apart did they discover how much they resembled motors in machines.

5 Note that, as explained early on, we make the case for departures from the machine metaphor by focusing on biological examples, much in the spirit of older discussions of mechanistic science. We do not make a claim to comprehensiveness, neither in the forms of departure from machine-style thinking nor in the types of examples we use.

6 There are important exceptions, perhaps the most obvious one being DNA. Transcription factors are an important exception too. See Milo et al. (2016, Chapter 2) for an extended (and highly illuminating) discussion. We look at issues related to low copy numbers later.

7 To understand how concentrations change as a result of feedback loops typically requires more than mental rehearsal of the mechanism. Instead, biologists often appeal to computational models. Bechtel and Abrahamsen (2010) and Brigandt (2013) describe how computational models often figure in explanations of mechanisms and refer to such accounts as "dynamic mechanistic explanations." Levy (2013) shows how Hodgkin and Huxley's (1952) explanation of the action potential employs differential equations to characterize the changing concentration gradients across the cell membrane. There is not space in this chapter to address the important role played by computational models.

8 The system includes a protein, CAP, that ensures that transcription from the lac operon happens only when glucose is not present. We set this aspect aside.

9 Some biologists have argued that such populational variation is adaptive (Dekel and Alon 2005; Kussell and Leibler 2005). We do not make any assumption on this score.

10  A referee notes that the appeal to robustness is unnecessary in this context: the case can be construed as one in which a choice between the models is based simply on the fact that one predicts that bursting depends on, e.g., ribosome copy number, whereas the other does not, and the latter prediction is correct. In a sense, this is so. But, first, the prediction in question is usually understood as a case of robustness, i.e., as a case in which a cellular (or more generally biological) function is performed despite external perturbations. Second, the determination of robustness is made theoretically, and it appears (although it is hard to show this for certain) that there is a theoretical bias in this case, as in others, in favor of robust models, presumably because of a background assumption that many biological systems are robust. For more on robustness as a criterion for model selection, see Green et al. (2015).

11  As with the earlier examples, the mechanism involves multiple copies of each molecule, and at any given time some molecules are in each of the phosphorylation states. Moreover, as in the following examples, these molecules are being synthesized as the mechanism operates (with the time of synthesis actually determined by the clock mechanism itself). To emphasize the key point of reorganization, we do not develop these other features.

12  Data about protein-protein interactions can be collected in high-throughput applications of the yeast two-hybrid technique and are often stored in online databases that network models access from platforms such as Cytoscape to generate a network representation into which additional data can be imported (see Bechtel, 2020).

## References

Abrahamsen, A., Sheredos, B., & Bechtel, W. (2018). Explaining visually using mechanism diagrams. In S. Glennan & P. Illari (eds.), *Routledge Handbook of Mechanisms*, 238–254. London: Routledge.

Bandyopadhyay, S., Mehta, M., Kuo, D., Sung, M. K., Chuang, R. et al. (2010). Rewiring of genetic networks in response to DNA damage. *Science*, 330, 1385–1389.

Bechtel, W. (2020). Data journeys beyond databases in systems biology. In S. Leonelli & N. Tempini (eds.), *Data Journeys in the Sciences*. Cham, Switzerland: Springer.

Bechtel, W., & Abrahamsen, A. (2005). Explanation: A mechanist alternative. *Studies in History and Philosophy of Biological and Biomedical Sciences*, 36, 421–441.

Bechtel, W., & Abrahamsen, A. (2010). Dynamic mechanistic explanation: Computational modeling of circadian rhythms as an exemplar for cognitive science. *Studies in History and Philosophy of Science Part A*, 41, 321–333.

Bechtel, W., Abrahamsen, A., & Sheredos, B. (2018). Using diagrams to reason about biological mechanisms. In P. Chapman, G. Stapleton, A. Moktefi, S. Perez-Kriz, & F. Bellucci (eds.), *Diagrammatic Representation and Inference*, 264–279. Cham, Switzerland: Springer International Publishing.

Bechtel, W., & Richardson, R. C. (1993/2010). *Discovering Complexity: Decomposition and Localization as Strategies in Scientific Research*. Cambridge, MA: MIT Press. edition published by Princeton University Press.

Brigandt, I. (2013). Systems biology and the integration of mechanistic explanation and mathematical explanation. *Studies in History and Philosophy of Biological and Biomedical Sciences*, 44, 477–492.

Choi, P. J., Cai, L., Frieda, K., & Xie, S. (2008). A stochastic single-molecule event triggers phenotype switching of a bacterial cell. *Science*, 322, 442–446.

Craver, C. F., & Darden, L. (2013). *In Search of Mechanisms: Discoveries Across the Life Sciences*. Chicago: University of Chicago Press.

Darden, L. (2002). Strategies for discovering mechanisms: Schema instantiation, modular subassembly, forward/backward chaining. *Philosophy of Science*, 69, 354–365.

Darden, L., & Craver, C. (2002). Strategies in the interfield discovery of the mechanism of protein synthesis. *Studies in History and Philosophy of Biological and Biomedical Sciences*, 33, 1–28.

de Lichtenberg, U., Jensen, L. J., Brunak, S., & Bork, P. (2005). Dynamic complex formation during the yeast cell cycle. *Science*, 307, 724–727.

Dekel, E., & Alon, U. (2005). Optimality and evolutionary tuning of the expression level of a protein. *Nature*, 436, 588–592.

Glennan, S. (2017). *The New Mechanical Philosophy*. Oxford: Oxford University Press.

Glennan, S., & Illari, P. M. (eds.). (2018). *The Routledge Handbook of Mechanisms and Mechanical Philosophy*, 1st ed. New York: Routledge.

Green, S., Levy, A., & Bechtel, W. (2015). Design sans adaptation. *European Journal for Philosophy of Science*, 5, 15–29.

Hegarty, M. (1992). Mental animation: Inferring motion from static displays of mechanical systems. *Journal of Experimental Psychology: Learning, Memory, and Cognition*, 18, 1084–1102.

Hodgkin, A. L., & Huxley, A. F. (1952). A quantitative description of membrane current and its application to the conduction and excitation of nerve. *Journal of Physiology*, 117, 500–544.

Ideker, T., & Krogan, N. J. (2012). Differential network biology. *Molecular Systems Biology*, 8, 565.

Jacob, F., & Monod, J. (1961). Genetic regulatory systems in the synthesis of proteins. *Journal of Molecular Biology*, 3, 318–356.

Kussell, E., & Leibler, S. (2005). Phenotypic diversity, population growth, and information in fluctuating environments. *Science*, 309, 2075–2078.

Lenstra, T. L., Rodriguez, J., Chen, H., & Larson, D. R. (2016). Transcription dynamics in living cells. *Annual Review of Biophysics*, 45, 25–47.

Levy, A. (2013). What was Hodgkin and Huxley's achievement? *The British Journal for the Philosophy of Science*, 65, 469–492.

Levy, A. (2014). Machine-likeness and explanation by decomposition. *Philosophers Imprint*, 14, 1–15.

Machamer, P., Darden, L., & Craver, C. F. (2000). Thinking about mechanisms. *Philosophy of Science*, 67, 1–25.

Milo, R., Phillips, R., & Orme, N. (2016). *Cell Biology by the Numbers*. New York: Garland.

Nathan, M. J. (2014). Causation by concentration. *British Journal for the Philosophy of Science*, 65, 191–212.

Nicholson, D. J. (2012). The concept of mechanism in biology. *Studies in History and Philosophy of Biological and Biomedical Sciences*, 43, 152–163.

Nicholson, D. J. (2013). Organisms ≠ Machines. *Studies in History and Philosophy of Biological and Biomedical Sciences*, 44, 669–678.

Steitz, T. A. (2008). A structural understanding of the dynamic ribosome machine. *Nature Reviews Molecular Cell Biology*, 9, 242.

Xie, X. S., Choi, P. J., Li, G.-W., Lee, N. K., & Lia, G. (2008). Single-molecule approach to molecular biology in living bacterial cells. *Annual Review of Biophysics*, 37(1), 417–444.

# 6 Magnetized memories

## Analogies and templates in model transfer

*Tarja Knuuttila and Andrea Loettgers*

## Introduction

The title of the Bakerian 2001 lecture by David Sherrington, a renowned physicist, and the other author of the Sherrington–Kirkpatrick model, was "On magnets, microchips, memories and markets: the statistical physics of complex systems."[1] That the Sherrington–Kirkpatrick model of spin glasses (i.e., disordered magnets) should find applications in as distant fields as statistical physics, computer science, neural network theory, and financial markets is both outstanding and commonplace. Indeed, it serves as an epitome of the contemporary modelling practice, where the same function forms and equations and mathematical and computational methods are being transferred and recycled across the disciplinary boundaries. While philosophers of science have only recently started to address the interdisciplinary dynamics of such a modelling practice, it is far from a new phenomenon. The transfer of theoretical and formal tools from one area of physics to another, and from physics to other disciplines such as economics and biology, has marked many scientific breakthroughs in the 19th and early 20th centuries. More recently, engineering has had an increasing interdisciplinary influence on many fields, attested by, for example, the emergence of synthetic biology.

Two bodies of philosophical discussion, in particular, have addressed this distinctively interdisciplinary character of modelling; one has studied analogical reasoning and the other has focused on the role of various kinds of templates in model-building and theoretical transfer. While the interest in analogical reasoning dates back to at least the 20th century through the discussion of both physical and mathematical analogies (e.g., Hertz 1893/1962, Hesse 1966), the focus on templates is of much more recent origin (e.g., Humphreys 2004, 2019; Knuuttila and Loettgers 2011, 2016; Houkes and Zwart 2019). The two discussions have proceeded in parallel, largely separately from each other.[2] Philosophers and cognitive scientists studying analogical reasoning have addressed the material and formal analogies between different objects/systems, in different domains, licensing inferences from the source objects/systems to the target objects/systems. By contrast, the emphasis of the discussion of templates has targeted cross-disciplinary formal and computational devices that are detached from any particular objects, systems, or domains.

At the outset, then, the analogy- and template-based approaches seem to be addressing different kinds of things that is also manifest in their different takes on representation. In introducing the notion of a computational template, Humphreys explicitly turns away from representation to computation, while analogical inference seems wedded to similarities and thus to "representation as." But appearances can be deceptive. Our claim is that any more fully blown account of model transfer, or that of modelling more generally, needs both perspectives, although, of course, actual cases of model transfer may proceed without making use of template-based and analogical reasoning. Moreover, the analogy- and template-based approaches lend more credence to each other. Without the perspective of analogical reasoning, it seems difficult to explain what drives the transfer of formal and computational templates from one domain into another, especially in interdisciplinary contexts, given that various kinds of structures may exhibit the appearances of the target domain of interest. On the other hand, the template-based approach focuses on the application of tractable tools, a phenomenon driving the model-based science, providing thus a more encompassing vision of modelling than the analogy-based approach alone.

In examining how the analogy- and template-based approaches contribute to each other, we focus on the conceptual dimension of model transfer, which has so far been inadequately addressed by the template-based view in its emphasis on tractability. Such inattention to the conceptual side of model transfer is also characteristic of those accounts of analogical reasoning that concentrate on structure mapping between two domains. Moreover, analogy-based approaches have typically addressed the local level of particular source and target domains. The focus on the conceptual dimension of model transfer bridges the local and the global in focusing on how general conceptual ideas embedded in formal templates facilitate the application of those templates across different domains. This is the representing-as dimension of template transfer that draws it close to analogical reasoning, anticipating and directing actual model construction.

We will examine analogical reasoning and template transfer through a study of the application of a spin glass model to neural networks in neurosciences. The Ising model of ferromagnetism (Ising 1925) provided the basic template for the Sherrington–Kirkpatrick model of spin glasses that in turn contributed to the Hopfield model of the associative memory (1982). The formal similarities between these three models are striking, underlining the insights of the template-based approach – yet the conceptual side of these model transfers is equally important. The reason for the transferability of the formal template underlying the Ising model, as well as that of its variants, such as the Sherrington–Kirkpatrick model, is that these models provide modelers hypothetical systems with interesting theoretical properties. These properties have been conceptually rendered as phase transitions, critical points, and, most importantly, cooperative phenomena, and they can be ascribed to various kinds of systems that appear to display similar kinds of behavior. Such model transfers presume flexibility of the formal template and involve novel theoretical interpretations and even development of new methods.

In the following sections, we will first present a brief overview of the philosophical discussion of analogies and templates and then study the transfers between the Ising model, the Sherrington–Kirkpatrick model, and the Hopfield model. The concluding section discusses the lessons to be learned concerning the use of analogies and templates in model-based theoretical practice.

## Analogies and templates

### *Analogies*

In philosophy of science, analogical reasoning has often been discussed in the context of knowledge generation, being related to topics such as scientific discovery, theory development, and hypothesis formulation. While the heuristic role of analogies in the aforementioned activities has generally been recognized, philosophers have disagreed as to whether any general account of analogical reasoning could be formulated that would warrant analogical reasoning in general (see Bartha 2016, 12–18; Norton 2011). In this regard, the use of analogical reasoning in modelling seems instructive. Analogical reasoning provides scientific modelers a frequently utilized, powerful cognitive strategy for transferring concepts, formal structures, and methods from one field to another. That analogical transfer is such a common practice in scientific modelling seems to request an analysis going beyond the conventional philosophical divide between discovery and justification. We will suggest that combining the insights concerning the modelers' use of cross-disciplinary templates with those of analogical reasoning will importantly contribute to such an analysis.

A classic treatment of analogical inference in the context of scientific modelling was provided by Mary Hesse (e.g., 1966). Hesse's account is two-dimensional: she distinguishes between "horizontal" and "vertical" relations. "Horizontal relations" refer to similar or dissimilar properties of two domains. Hesse approaches them in terms of positive, negative, and neutral analogies. Positive analogies refer to those properties that the two analogs have in common, whereas negative analogies refer to known differences between them. Neutral analogies refer to the properties whose commonality or difference has yet to be established, thus providing epistemic potential for further inferences and theoretical development. "Vertical relations" are relations between objects and properties *within* a domain. Two domains are formally analogous if the relations between certain elements within one domain are identical or at least comparable to the relations of the corresponding elements in another domain. Hesse contrasts such formal analogies to material similarities that are shared, frequently observable, or pretheoretic resemblances between two domains. In her view, acceptable analogies need to be grounded in both horizontal and vertical relations; in her examples, material analogies seem to provide a basis for constructing formal analogies.

Cognitive scientist Dedre Gentner has made formal analogies the centerpiece of her influential theory of analogy (e.g., 1983). She distinguishes between attributes and relations, claiming that an analogy does not necessarily become stronger if the two analogs share only more attributes. According to Gentner, the key similarities are those that lie in the relations that hold within a domain, and it is preferably those relations that are being transferred to another domains. These kinds of formal analogies display systematicity, being governed by "higher order relations," such as causal, mathematical, or functional relationships. The fact that relations like these are being sought after in scientific reasoning underlines, in Gentner's view, the importance of formal analogies. She seems thus not to agree with Hesse, who stresses the importance of causal relations but does not contrast them to horizontal relations between observable properties. Given the overriding focus of Gentner's account on formal analogies, analogical transfer becomes that of structure-mapping between the source and target domains (Gentner 1983; see also Gentner and Markman 1997). With structure-mapping, Gentner refers to mapping knowledge about the base domain into the target domain, such that the mapping rules "depend only on the syntactic properties of the knowledge representation, and not on the specific content of the domains" (Gentner 1983, 155).

Yet analogical inference cannot boil down to a mere projection of a syntactic structure. Any system can be conceptualized in different ways, resulting in different kinds of structures. Consequently, the structure a system is supposed to exhibit is intimately related to how the system is described (see Frigg 2006). Indeed, in his discussion of analogical reasoning, Bartha (2016, 32–33) argues that there is "no short-cut via syntax." He underlines the importance of focusing on the *relevant* features of both domains and how they relate to the analogical inference in question.

Bartha's "articulation model" addresses relevance by paying particular attention to "prior association" and "potential for generalization" (Bartha 2010). Prior association holds in the source domain between the known similarities and a further property that is then projected on the target domain. It picks up the features of the source that are deemed relevant for the analogical inference. Bartha also underlines the need to make the prior association explicit. The potential for generalization, in turn, stipulates that there must be good grounds to believe that the same kind of connection would hold in the target domain. In particular, there should be no critical disanalogies between the domains. For instance, in analogical transfer between the Sherrington–Kirkpatrick model and the Hopfield model, the prior association between the interactions between the magnetic moments and cooperative phenomena, i.e., ferromagnetism, is generalized to hold also in the case of pattern matching performed by the nervous system.

What we want to focus our attention on, then, is the fact that even in the case of formal analogies, the projection of a structure is never purely syntactic. The point is that the templates used in model construction come with associated concepts that aim to capture some theoretically interesting properties of the structures in question. These concepts suggest how to theorize about the phenomenon of interest and function as springboards for further theoretical development in the target domains. Accordingly, it is possible to extend Bartha's requirement of the potential for generalization from the nonexistence of critical disanalogies to the utilization of some general theoretical principles. These principles are embedded in well-understood formal and computational templates that are used to study multiple phenomena. We will examine this insight in more detail in the next section by discussing templates of various kinds.

### Templates

The interest in templates as distinctively cross-disciplinary vehicles of modelling has its origin in the work of Paul Humphreys (e.g., 2004, 2019). He called attention to one of the most conspicuous characteristics of the contemporary modelling practice: its reliance on "the relatively small number of computational templates in the quantitatively oriented sciences" (2004, 68). "Science," he suggested, "would be vastly more difficult if each distinct phenomenon had a different mathematical representation" (ibid.). This observation, we suggest, is even more true of model-based science.[3]

Humphreys zooms in on something so "simple and well-known" (2004, 60) that it has escaped the explicit attention of philosophers of science: computational templates. Humphreys' computational templates are genuinely cross-disciplinary mathematical or computational forms and methods that can be applied to different problems in various disciplines. In using the word "template," Humphreys invokes a pattern for developing a product that can simultaneously be configured in view of the aims of the modeler (2019, 5).

Computational templates may have their origin in formal disciplines, such as the Poisson distribution in probability theory. While computational templates are genuinely subject-independent in their application, many of them were originally introduced as theoretical models of a certain system, being only subsequently applied to different domains. A theoretical model functions as a template first when it is separated from the original theoretical context and used to model other, usually very different, types of phenomena. The Lotka-Volterra model[4] and the Ising model provide illustrative examples

of templates that are so general they have been used in virtually all areas of science where scientists are engaged in mathematical model-building. But apart from generality, to allow for computation, computational templates need to be tractable. Humphreys (2004) considers tractability to be the distinguishing feature of computational templates. Interestingly, many models, such as the Lotka-Volterra model and the Ising model, have gained such tractability due to the subsequent development of mathematical and computational methods. Generic technologies such as digital computing made the Lotka-Volterra model a suitable case for studying the dynamics of nonlinear systems (Knuuttila and Loettgers 2011). By contrast, mathematical methods, developed within physics, such as the renormalization group theory, played a crucial role in making the Ising model tractable.

Humphreys (2019) seeks to distinguish the template-based approach from analogical reasoning. His main argument is that while analogical reasoning relies on similarities typically left at least partially implicit, the template-based approach does not need to rely on any "vague" resemblances. Instead, theoretical templates that are candidates for developing into transdomain templates are typically results of construction processes[5] whose assumptions can be made explicit. As a consequence, Humphreys claims, there are cases where "there is no need to use analogical reasoning in applying a template – we can check directly whether the assumptions are satisfied for the system at hand" (Humphreys 2019, 115). Moreover, Humphreys (2019) notes, in line with his earlier writings on templates, that any transfer of a template usually involves refinement and adaptation of the template to a new domain, except for the off-the-shelf representations "that can be opportunistically justified at the system level by analogical reasoning from their previous successful applications that are recognized as similar" (ibid.). According to Humphreys' examples, statistical distributions as well as many general equation forms belong to this group. However, when it comes to the latter group, their off-the-shelf nature is questionable, as our case shows.

Another related question concerns the application of formal templates with only mathematical interpretation and whose "construction assumptions have only mathematical content" (Humphreys 2019, 114).[6] Humphreys seeks to explain how it is possible to apply formal templates given their purely mathematical interpretation. He approaches such an application as a mapping from a formal template to a target system. In such a mapping, all empirical content is contained within the mapping and not within the template, which explains why formal templates can be applied across a multitude of domains. Humphreys views the construction and use of formal templates as superior to analogical inference since in using them scientists do not need to invoke the language of the domain from which the template originally comes: "[t]here is therefore no need for vocabulary translations or for interdisciplinary knowledge" (Humphreys 2019, 118). He uses as an example the Barabási-Albert model (Barabási and Réka 1999), which is a random scale-free network making use of a preferential attachment mechanism. It has been applied to various kinds of natural and social networks that contain nodes ("hubs") whose number of links within a network greatly exceeds the average.

The Barabási-Albert model fits Humphreys' views on formal template transfer, as its origin is in mathematical theory and so it is devoid of previous subject-specific empirical and theoretical content. But does such a case of formal template transfer also suit other kinds of template transfers characteristic of contemporary modelling practice? And what explains the seemingly unreasonable success of a relatively small number of templates? We do not believe that their success can be explained by tractability and generality alone unless these two features are linked in a particular way. Specifically, successful templates embody something more: a vision of the phenomenon exhibiting a particular kind of *general* pattern for the study of which the template offers *tractable* or at least already well-studied tools. And seeing various kinds of systems as instances of some already familiar general patterns amounts – to use Kuhnian language – to a "gestalt switch" that enables scientists to approach various kinds of systems as being like each other – at least in one important dimension.[7] To study this analogous moment in template transfer, we turn to our case study on the model transfer between the Sherrington–Kirkpatrick model and the Hopfield model. The idea of cooperative phenomena forms the conceptual core of this template transfer, already introduced by the Ising model, which provided some basic formal templates and associated theoretical ideas for the Sherrington–Kirkpatrick and Hopfield models.

## Modelling cooperative phenomena

The Ising model, originally presented as a mathematical model of ferromagnetism by Ernst Ising (1925), is nowadays used to study an amazing variety of phenomena in different disciplines, ranging from physics to biology and social sciences. Although, in the case of the Ising model, what appear to be transmitted between different fields are mathematical structures, the conceptual side of these model transfers has been equally important. Physicist Daniel Amit has described the conceptual fruitfulness of the Ising model in physics the following way:

> [The Ising model] has been a birthplace and the testing ground for a treasure of new concepts in essentially all fields of physics. Such fundamental ideas as symmetry breaking, cooperative phenomena, order parameters, disorder parameters, critical exponents, symmetry restoration etc., have had their first explicit, precise articulation in the framework of this apparently simple, naïve model.
>
> (Amit 1989, 105)

While most of these theoretical ideas have been developed and applied in the domain of physics, the concept of "cooperative phenomena" has especially proved globally applicable, being applied beyond physics in biology, economics, and sociology. How did cooperative phenomena gain this global, cross-disciplinary nature? In the following section, we will trace the journey of this concept from the context of modelling properties of magnetic systems in physics to neuroscience.

### *The Ising model*

Cooperative phenomena are general in character, resulting from interactions between the constituents of a system. These interactions can be of various kinds. Ferromagnetism provides a standard example that is also of historical importance. On the microscopic level, a piece of magnetic iron consists of magnetic moments, which below a certain temperature $T_C$ align and result in a macroscopic net magnetization. Above the temperature $T_C$, the thermal motion of the magnetic moments counterfeits this tendency and, as a result, the net magnetization vanishes and the piece of iron becomes paramagnetic. The transition from the ferromagnetic phase into the paramagnetic phase (and the other way around) is called "phase transition" and $T_C$ the "critical temperature." This kind of transition can be observed in experiments on the macro level, but the interactions between magnetic moments on the micro level that give rise to the transition are not experimentally accessible. The Ising model provides a conceptual and methodological framework by which these processes on the micro level can be approached.

At first glance, the structure of the Ising model seems astonishingly simple for such a consequential model. It consists of $N$ magnetic moments, so-called spins $S_i$, which can take only two values, $S_i = +1$ or $-1$, corresponding to their two possible discrete orientations up and down. In the two-dimensional case, the spins are located on the sides of a lattice. The interaction, which is central for the occurrence of cooperative phenomena such as ferromagnetism, is given by the interaction energy $J_{ij}$, describing the interaction strength between the *nearest neighbor* spins $S_i$ and $S_j$. In the original Ising model, $J_{ij}$ is constant, and in the case of ferromagnetism, where all the spins are aligned, the interaction energy is positive ($J > 0$). To sum up, the interaction energy depends on the configurations of neighboring spins and, furthermore, tends to align them.

With each magnetic moment $S_i$ comes an internal magnetic field $h_i$ that is created by the interaction between the magnetic moments $S_i$:

$$h_i = \sum_{j, j \neq i}^{N} J_{ij} S_j \tag{1}$$

with $J_{ij} = J_{ji}$.

For each of the $2^N$ configurations of spins $\{S\}$, where 2 stands for the two possible orientations of the spins and $N$ for the total number of spins, an energy function is given for each of these microstates by:

$$E\{S\} = -\frac{1}{2} \sum_{i, j \neq i}^{N} J_{ij} S_i S_j. \tag{2}$$

The overall energy of the system decreases if $S_i$ and $S_j$ point in the same direction. In this case, the interaction energy $J_{ij}$ is making a positive contribution, which, together with the " $-$ " sign in front of the sum, leads to the decrease of the overall energy $E$. If and $S_j$ point in different directions, the overall energy of the system increases because the interaction energy $J_{ij}$ is making a negative contribution.

The formal structure of the magnetic moments $S_i$, the internal magnetic field $h_i$, and the energy $E\{S\}$ provide conceptual and methodological resources by which cooperative phenomena can be explored. In Humphreys's terms, they can be approached as templates. One of the main challenges consists in calculating from the microscopic behavior the macroscopic properties, such as the magnetization $M$. In general, calculating some macroscopic property from the microstates of the system is one of the main subjects of statistical mechanics. The Ising model provided one successful framework for such a task, developed in the confined context of ferromagnetic systems.[8]

In statistical mechanics, the magnetization $M$ is calculated from the microstates $\{S\}$ in the following way:

$$\langle M \rangle = \frac{\sum\limits_{\{S\}} M\{S\} e^{-E/kT}}{\sum\limits_{\{S\}} e^{-E/kT}} \tag{3}$$

In the equation, the sum $\sum_{\{S\}} e^{-E/kT}$ is the so-called partition function. The functional form of the energy $E$ given by equation (2) is specific to the Ising model due to its form of interaction. The exponential function describes the probability of the realization of a microstate at a given energy and temperature. Because of the minus sign in the exponential, microstates at a high energy are less probable than states at a lower energy. The probability that a microstate is realized can be calculated by:

$$P\{S\} = \frac{e^{-E/kT}}{\sum_{\{S\}} e^{-E/kT}}$$

Accordingly, the magnetization is the sum over all $2^N$ possible states weighted by the probability that a microstate is taken. The partition function and energy function can be considered as general theoretical templates in the context of physics. The partition function has its origin in statistical mechanics, whereas the energy function is fundamental in all parts of physics. The probability distribution provides a computational template. Those templates are flexible enough for modelling different forms of interaction and therefore cooperative behavior. The package consisting of the concept of cooperative phenomena and the associated theoretical and computational templates served as a framework within which further concepts, such as phase transitions, critical exponents, and symmetry breaking, got developed, together with further methods for their calculation. Within physics, these concepts and methods have not subsequently been confined to the case of ferromagnetism. They have also been applied to other physical systems exhibiting cooperative phenomena. The so-called spin glasses provide one particularly fruitful application. Spin glasses differ from ordinary glasses by containing a small number of magnetic moments interacting with each other. These interactions lead to interesting cooperative behavior due to the fact that both ferromagnetic and antiferromagnetic couplings are present in the system. The simultaneous presence of ferromagnetic and antiferromagnetic couplings, in general, does not allow for the establishment of a conventional long-range order (of ferromagnetic or antiferromagnetic[9] type).

### *The Sherrington–Kirkpatrick model*

In 1978, David Sherrington and Scott Kirkpatrick introduced a model of a spin glass by drawing an analogy to the case of ferromagnetism (Sherrington and Kirkpatrick 1978). They hypothesized that the observed behavior of spin glasses is caused by the interaction between their magnetic moments, as is the case in the Ising model. The subsequent development of the Sherrington–Kirkpatrick model (hereafter the SK model) made available further concepts, templates, and methods.[10] The flexibility of the theoretical and formal templates underlying the Ising model allowed for the construction of the SK model, which tries to capture spin glass–specific observations, such as a transition into a disordered state at low temperatures. However, at first glance the SK model cannot be distinguished from the Ising model.

As in the Ising model, the magnetic moments in the SK model are represented by binary variables $S_i$. Again, the magnetic moments $S_i$ can take either the value +1 or -1. The coupling between two spins, $S_i$ and $S_i$, is, as before, in the case of the Ising model, represented by the coefficient $J_{ij}$, and the overall energy of the system is also of the form:

$$E = -\sum_{i,j \neq i} J_{ij} S_i S_j$$

The main difference between the two models lies in the form of interaction. In the SK model, the couplings are modeled as a function of the distance between the magnetic moments $J_{ij}=J(R_i-R_j)$, with $R_i$ and $R_j$ as the positions of the magnetic moments on, for example, a lattice. This kind of interaction leads to cooperative behavior, which is different than in the case of the Ising model. Positive values of $J_{ij}$ correspond to ferromagnetic and negative values to antiferromagnetic couplings. The spins in this model cannot at the same time satisfy both ferromagnetic and antiferromagnetic couplings, which means that the couplings are of a competitive nature. The consequences of these competing interactions between the ferromagnetic and antiferromagnetic couplings become apparent at low temperatures, when the system undergoes something like a phase transition. The system exhibits a "freezing transition" to a state with a new kind of order in which the magnetic moments are aligned in random directions (Binder and Young 1986). The topology of the energy landscape of spin glasses after undergoing this freezing transition is varied, consisting of a large number of valleys, representing metastable or stable spin configurations.

The flexibility of the partition and energy functions allowed scientists to explore what kinds of macroscopic properties, such as the freezing transition, were caused by the microscopic behavior. These calculations turned out to be very difficult and required the development of further mathematical tools, such as the replica method, which is used in statistical mechanics in the calculation of the partition function (see footnote 12).

The Ising and SK models display how the concept of cooperative phenomena coupled with theoretical and formal templates and computational methods accounts for macrolevel phenomena, such as phase transitions, in terms of microlevel interactions. Such cooperative phenomena are not limited to the physical sciences. The question is how template transfer from physics to other sciences, licensed by the notion of cooperative phenomena, is bound to succeed. An example of such a transfer is provided by the Hopfield model, which transferred the SK model of spin glasses to neuroscience. While the SK model successfully applied the Ising model to spin glasses, explaining some characteristics of spin glasses via their ferromagnetic and antiferromagnetic moments, any such straightforward interpretation of microlevel phenomena was not possible in neuroscience. As a result, analogical reasoning played a much more substantive role in the case of the Hopfield model.

### The Hopfield model

Theoretical physicist John Hopfield introduced in 1982 a model that was to become one of the milestones in the study of artificial neural networks (Hopfield 1982). He was interested in how the brain could fulfill the task of completing information, such as recognizing a friend's face, from an incomplete input (i.e., the partial picture provided by our visual perception). Hopfield made use of both positive and negative analogies in his reasoning. First, he drew a negative analogy to small circuits, such as electric circuits and computers, arguing that evolution does not proceed in the same way as an engineer. Second, he drew a positive analogy to many body systems, suggesting that auto-associative memory could be understood as collective[11] phenomena:

> Given the dynamical electrochemical properties of neurons and their intercon-
> nections (synapses), we readily understand schemes that use a few neurons to
> obtain elementary useful biological behavior. Our understanding of such sim-
> ple circuits in electronics allows us to plan larger and more complex circuits,
> which are essential to large computers. Because evolution has no such plan,
> it becomes relevant to ask whether the ability of large collections of neurons
> to perform "computational" tasks may in part be a spontaneous collective
> consequence of having a large number of interacting simple neurons.
>
> (Hopfield 1982, 2554)

The turn toward collective phenomena does not, by itself, give too much understanding of how the brain possibly recognizes a pattern from an incomplete input. In the construction of the actual model, Hopfield began by visualizing the process of pattern recognition as water flowing from different directions into a valley. Take again the example of the friend's face. In addition to her, we memorize a lot of other different people or objects. Translated into the visual analogy of Hopfield, such phenomena amount to a landscape consisting of many valleys in which each valley stands for one memorized person and object. This valley analogy renders the problem of pattern recognition as that of finding how patterns (valleys) are stored and the dynamic by which an incomplete pattern is recognized and, by so doing, completed.

The complex structure of the landscape of the SK model, consisting of many energy minima, provided a suitable template for Hopfield's idea of how pattern recognition should be approached.[12] The notion of energy minima, which was one of the central theoretical elements transferred from the SK model, was interpreted by Hopfield in terms of stored patterns. In his analogical reasoning, Hopfield modified and integrated different templates, such as the energy function; rendered the neural components as binary variables; and introduced dynamic and storage rules from statistical mechanics and neuroscience. The construction of the Hopfield model was far from any straightforward application of the SK model.

One challenge was due to the randomness of the energy minima in the SK model. This cannot be the case if energy minima stand for a stored pattern to be recovered by the recognition system. To accommodate this feature, Hopfield made use of the Hebb learning rule (Hebb 1949). According to this rule, the synaptic efficiencies between neurons are described by the set of parameters $J_{ij}$ in which the information is stored. The simultaneous activation of two connected neurons results in a strengthening of the synaptic coupling between the two neurons. This rule is formalized in the Hopfield model as follows:

$$J_{ij} = \sum_{\mu=1}^{P} \xi_i^{\mu} \xi_j^{\mu}$$

The $\xi_i^{\mu}$ are variables that describe a pattern, i.e., a given configuration of active and inactive neurons. The number of patterns stored in the network is given by p, and in each pattern the number of neurons is equal to the total number of neurons in the network N. Each of the patterns is associated with an energy minimum. The topology of the energy landscape shows a similar complexity as that displayed by the SK model. By implementing the Hebb rule, which even has some neurophysiological grounding,[13] the patterns are not random anymore. On the formal level, the structure of the Hopfield model is akin to the Ising and SK models. The main difference between them – as is also the case between the Ising and SK models – lies in the choice of the coupling between the components of the network.

As in the case of the Ising and SK models, the neurons $\sigma_i$ in the Hopfield model are binary variables. The neuron $\sigma_i$ takes the value 1 in case it is active and the value 0 if it is inactive. The state of each of the neurons is determined by its post-synaptic potential (PSP)*hi*, produced by the activating signals arriving from all the other neurons to which it is connected. It is given by:

$$h_i = \sum_{j,j \neq i}^{N} J_{ij} \sigma_j$$

The PSP is of the same form as the internal magnetic field caused by the magnetic moments, although it has a different interpretation in the case of the Hopfield model. The internal magnetic field in the Ising model is the magnetic field that the magnetic moment $S_i$ experiences. It either aligns the two magnetic moments or lets them point in different directions. In the case of the Hopfield model, the magnetic field is replaced by a biochemical interaction that either changes the state of the neuron, e.g., from active to inactive, or leaves it in its actual state. Moreover, the energy function is of the same form as in the Ising and SK models:

$$E = -\frac{1}{2} \sum_{i \neq j} J_{ij} \sigma_i \sigma_j$$

It assigns an energy value $E$ to each system configuration $\sigma = \{\sigma_1, ...., \sigma_N\}$. The energy function can be considered as a computational template that is adjustable to the respective system to be modeled. A further important difference between the Ising, SK, and Hopfield models can be seen in their dynamics. The Ising and SK models are not dynamic; they calculate the properties of the system, such as the magnetization of the microstates, through the probability of their occurrence. By contrast, dynamics are essential for the Hopfield model: starting from an incomplete input, the neural network develops into an energy minimum associated with one of the stored patterns.

To conclude, the Hopfield model provides an example of how the notion of collective phenomena enabled Hopfield to draw an analogy between auto-associative memory and the phenomena modeled by the Ising model and the SK model. This analogy enabled Hopfield to transfer theoretical and computational templates from the study of magnetic phenomena into the field of artificial neural networks. These templates functioned as conceptual and methodological resources, which were used to construct an artificial neural network that was able to recognize patterns.

## Analogies and model templates in model transfer

We have shown that what made the Ising model and its offspring, like the Sherrington–Kirkpatrick model, attractive candidates for model transfer is the conceptual and methodological framework they embody. This framework renders certain kinds of patterns as instances of cooperative phenomena coupled with associated mathematical forms and tools that enable the study of such phenomena. The Sherrington–Kirkpatrick model examines a situation where the behavior of magnetic spins is disordered due to competing ferromagnetic and antiferromagnetic couplings between the magnetic moments. This situation leads to behavior that cannot be anticipated from any single element of the system but only from the competing interactions between a large number of individual elements, leading to a large number of local minima. In the course of cooling down, the spin glass gets trapped into one of the many local minima of the complex energy landscape. Hopfield was able to use this property in modelling auto-associative memory.

The notions of either a computational template or a formal template do not adequately recognize this intertwinement of the conceptual, mathematical, and computational sides of model transfer.

To better capture the holistic aspect of modelling, Knuuttila and Loettgers introduced the notion of a "model template" that is a mathematical structure or computational method that is "coupled with a general conceptual idea that is capable of taking on various kinds of interpretations in view of empirically observed patterns in materially different systems" (2016, 396). As such, a model template provides "a formal platform for minimal model construction coupled with very general conceptualization without yet any subject-specific interpretation or adjustment" (ibid., 382). The Ising model provided such a model template for the SK model, and the SK model, in turn, provided a model template for the Hopfield model. This model template can be understood as a formally defined framework for modelling particular kinds of cooperative systems that instantiates the concept of cooperativity through the interaction energy $J_{ij}$, [14] i.e., the coupling strength between the magnetic moments of the system. The interaction energy is central for the cooperative behavior of the system. It defines the form of the energy landscape by being embedded in the interlocking theoretical templates of energy function $E\{S\}$, magnetization $\langle M \rangle$, and partition function $P\{S\}$. Transferred to the Hopfield model, these templates forgo their original interpretation, becoming thus computational templates. [15]

Thinking about the SK model as a model template for the Hopfield model emphasizes the importance of the analogical dimension of template transfer. The notion of a cooperative mechanism provides the central conceptual idea shared by the Ising, SK, and Hopfield models. That Hopfield was able to conceive of pattern recognition in terms of the energy landscape resulting from competing ferromagnetic and antiferromagnetic couplings between the magnetic moments was a result of analogical reasoning and not of any unequivocal structure mapping. Moreover, the analogy enabled him to make use of the theoretical and computational templates provided by the Ising model and the SK model, leading to an intricate model construction process in which Hopfield also drew resources from statistical mechanics.

According to Humphreys, one can often dispose of analogical reasoning because the model construction assumptions can be stated explicitly and checked empirically. This may hold true in some cases in physics but not in the case of the SK model, which does not lend itself to any straightforward empirical interpretation. And, in the case of transdomain transfer, checking empirically the assumptions of the model may even be more difficult. [16] In the case of the Hopfield model, it is difficult to see how the construction assumptions could be checked, as the concepts adopted from physics, such as temperature or phase transitions, do not map onto any empirical properties of neural networks.

In our view, analogy- and template-based approaches can fruitfully be used to augment each other. In analogy-based approaches, the formal and mathematical representations are often considered to be derived by abstraction from target and source domains. Analogy then enables the mathematization of the target domain in terms of relational generalizations, which may yield abstract schemas. For instance, Nersessian (2002) details how Maxwell formulated the mathematical representation of the electromagnetic field concept by making use of imaginary models of fluid medium, drawing, moreover, from continuum mechanics and machine mechanics. As he progressed in this theorizing, his conception of the aetherial medium became more abstract. Nersessian's discussion captures the conceptual and intertheoretical dimension of analogical exchange but also displays the tendency of the analogy-based approaches to disregard the genuinely cross-disciplinary nature of many formal tools.

By contrast, the template-based approach focuses on the generalized methods for modelling various kinds of systems. It also addresses the question of why some tractable templates have proven so nearly universally applicable. Apart from mentioning their generality and tractability, Humphreys underlines the importance of the local construction and adjustment of templates. We have suggested yet another reason for this universality, highlighting the importance of the analogy-based approach: crucial for template transfer is the general conceptual core of the model template. This conceptual core is global in character, motivating local and domain-specific template construction and adjustment processes. While the templates themselves may appear to be merely syntactic structures in transdomain exchange – given that their underlying ontologies change with the different material systems they are applied to – they do also have an important conceptual dimension.[17] And this is animated by analogies between various kinds of systems that are used to mobilize template transfers across a wide spectrum of domains and disciplines, a practice that is particularly visible in contemporary complex systems theory and network science.

Last, and related to the global character of model templates, we wish to briefly consider how Hopfield himself understood model transfer from physics to neuroscience. In a talk on the work of Sir David McKay, Hopfield referred to Niels Bohr and Max Delbrück, who both thought that to describe and explain biological phenomena, a new kind of physics would be necessary. They asked "[h]ow the diverse seemingly purposeful complex phenomena described by the word 'life' could emerge from lifeless physics."[18] Hopfield argued that from the present-day vantage point, the idea of a new kind of physics may have become obsolete with the realization that the laws of neural-based behavior in higher animals are macroscopic. Biological and large physical systems are alike, in that both have robust emergent properties arising from the interaction between the components of the system. Herein lies the justification for Hopfield for drawing an analogy between magnetic systems and neural networks. The analogy is based on the shared structural and dynamical properties of systems giving rise to specific properties such as ferromagnetism or pattern recognition. It is not justified on the basis of any observed fact of analogy between particular systems (cf. Norton 2011). The justification is more general and theoretical in nature and related to a new grouping of systems and phenomena under the heading of many body systems as well as emergent behavior that enables the analogical transfer of theoretical and model templates within physics but also beyond physics to biology.[19]

## Notes

1 The Bakerian Medal and Lecture is awarded annually by The Royal Society, and it is one of the most prestigious lectures in physical sciences.
2 See, however, Humphreys (2019).
3 Paul Humphreys originally introduced the notion of a computational template in his study of computer simulations and their relation to traditional modelling techniques (see also Humphreys 2019, 3).
4 See Houkes and Zwart (2019) for a study of the application of the Lotka-Volterra model to technology transfer. The model was originally used by Volterra to study population dynamics and Lotka to study biological and chemical systems more generally (Knuuttila and Loettgers 2011).

5  Humphreys contrasts constructed transdomain templates to those theoretical templates that are part of the fundamental principles of a theory, such as Newton's second law or the Schrödinger equation.

6  Humphreys (2004) does not specifically address formal templates.

7  There is an analogous moment even in the transfer of what seems a purely formal template. Such an analogous dimension of model transfer depends, we suggest, on the conceptualization of the phenomenon as being of a particular kind and thus giving rise to some distinctive patterns.

8  Another example is provided by the kinetic gas model.

9  In the antiferromagnet, neighboring spins point in different directions. In the paramagnetic phase, due to the temperature, the spins point in random directions.

10  Probably the most important method developed in this context is the replica method, which allows for the calculation of the sum over the $2^N$ microstates, which easily becomes a very large number. The different possible realizations of disorder also pose a serious problem: the form of phase transitions varies depending on the distribution of the interactions between the magnetic moments. This means that there exists a correlation between disorder and the form of the phase transition. To get more representative results, an average of a large number of different realizations of interactions – replicas – is made use of (see Mezard et al. 1987).

11  Hopfield used the word "collective" synonymously with what we call "cooperative."

12  Hopfield was not the first one to draw an analogy between the Ising model and the organization of neurons (see Cragg and Temperley 1954; Caianiello 1960).

13  Direct experiments on neurons have shown that changes in the signaling transfer are part of learning in the brain (e.g., Yang et al. 2014).

14  Interaction energy is a general concept that can be found in physics, chemistry, and biology.

15  In the Hopfield model, the equations for modelling the dynamics of pattern recognition, i.e., Glauber dynamics, are more generally used to model the stochastic dynamics in the Ising model. Glauber dynamics can be compared to computational templates such as the Poisson distribution. A question that remains unanswered by Humphreys is that of the relationship between computational and formal templates. The notion of a computational template has receded in the background in Humphreys (2019). Poisson distributions are in Humphreys' (2004) examples of computational templates, but they are discussed as formal templates in Humphreys (2019). Does this mean that the notion of a formal template covers computational templates? And how, then, are computational templates related to formal templates such as Barabási-Albert preferential attachment templates? How these lines are drawn does not seem to be of consequence to our argument since, in addition to the formal or computational side of template transfer, the notion of a model template also encompasses a conceptual dimension.

16  Humphreys does not deny that analogical reasoning may play some role, as shown by his discussion of the application of Volterra's predator-prey model to the dynamical contradictions of capitalism (Goodwin 1967). By referring to the idea of symbiosis between two populations that are partly complementary and partly hostile, he grants that "Kuhnian analogies can assist in the transfer of a representation from one domain to another" (Humphreys 2019, 116). Yet he insists that "this analogical transfer can be made explicit by means of a formal set of assumptions" (ibid); in the case of the Goodwin model, the Lotka-Volterra equations can be arrived at via explicit economic assumptions. Moreover, in Humphreys' view, formal templates, such as Barabási networks, are instantiated by the mapping of a formal template on a target system, and the success of this mapping process is evaluated in terms of whether the formal construction assumptions are empirically justified. Thus, Humphreys appears to argue that analogies can be dispensed with or that they are not needed in the first place.

17  Although Humphreys (2004) argues that templates are endowed with intended interpretation, he also mentions that changing the ontology of the system comes "very close

to starting a new [template] construction process" (Humphreys 2004, 80). If the template is used to model a new system, the original justification goes with the intended interpretation (ibid.). What we are arguing is that the general conceptual content of a model template still bestows it with some justification – along with tractability – and the interesting philosophical question is what kind of justification this is.

18 Inference, Information, and Energy: A Symposium to celebrate the work of Professor Sir David McKay. University of Cambridge 15.3.2016. Hopfield had been McKay's PhD supervisor.

19 This project has received funding from the European Research Council (ERC) under the European Union's Horizon 2020 research and innovation programme (grant agreement No 818772).

## References

Amit, D. (1989). *Modeling Brain Function: The World of Attractor Neural Networks*. Cambridge, MA: Cambridge University Press.

Barabási, A-L., & Réka, A. (1999). Emergence of scaling in random networks. *Science*, 286, 509–512.

Bartha, P. (2010). *By Parallel Reasoning: The Construction and Evaluation of Analogical Arguments*. New York: Oxford University Press.

Bartha, P. (2016). Analogy and analogical reasoning: Winter 2016 edition. In *Stanford Encyclopedia of Philosophy*. Stanford, CA: Center for the Study of Language and Information Stanford University.

Binder, K., & Young, A. P. (1986). Spin glasses: Experimental facts, theoretical concepts, and open questions. *Reviews of Modern Physics*, 58(4), 801–976.

Caianiello, E. R. (1960). Outline of a theory of thought-processes and thinking machines. *Journal of Theoretical Biology*, 1(2), 204–235.

Cragg, B. G., & Temperley, H. H. (1954). The organization of neurons: A cooperative analogy. *Electroencephalography and Clinical Neurophysiology*, 6(1), 85–92.

Frigg, R. (2006). Scientific representation and the semantic view of theories. *Theoria*, 55, 49–65.

Gentner, D. (1983). Structure-mapping: A theoretical framework for analogy. *Cognitive Science*, 7, 155–170.

Gentner, D., & Markman, A. B. (1997). Structure mapping in analogy and similarity. *American Psychologist*, 52(1), 45–56.

Goodwin, R. M. (1967). A growth cycle. In C. H. Feinstein (ed.), *Socialism, Capitalism and Economic Growth* (Essays presented to Maurice Dobb). Cambridge: Cambridge University Press.

Hebb, D. (1949). *The Organization of Behavior: A Neurophysiological Theory*. Mahwah: Erlbaum Books.

Hertz, H. ([1893] 1962). *Electric Waves: Being Researches on the Propagation of Electric Action with Finite Velocity Through Space*. New York: Dover Publications.

Hesse, M. (1966). *Models and Analogies in Science*. Notre Dame: Notre Dame University Press.

Hopfield, J. (1982). Neural networks and physical system with emergent collective computational abilities. *Proceedings of the National Academy of Sciences of the USA*, 79(8), 2554–2558.

Houkes, W., & Zwart, S. D. (2019). Transfer and templates in scientific modelling. *Studies in History and Philosophy of Science* 77, 93–100.

Humphreys, P. (2004). *Extending Ourselves. Computational Science, Empiricism and Scientific Method*. Oxford: Oxford University Press.

Humphreys, P. (2019). Knowledge transfer across scientific disciplines. *Studies in History and Philosophy of Science*, 77, 112–119.

Ising, E. (1925). A contribution to the theory of ferromagnetism. *Zeitschrift für Physik*, 31(1), 253–258.

Knuuttila, T., & Loettgers, A. (2011). The productive tension: Mechanisms vs. Templates in modeling the phenomena. In P. Humphreys & C. Imbert (eds.), *Representations, Models, and Simulations*, 3–24. New York: Routledge.

Knuuttila, T., & Loettgers, A. (2016). Model templates within and between disciplines: From magnets to gases – and socio-economic systems. *European Journal for Philosophy of Science*, 6(3), 377–400.

Mezard, M., Parisi, G., & Virasoro, M. A. (1987). Spin glass theory and beyond: An introduction to the replica method and its applications. *World Scientific Lecture Notes in Physics*, Vol. 9. Singapore: World Scientific Publishing.

Nersessian, N. (2002). Maxwell and the method of physical analogy: Model-based reasoning, generic abstraction, and conceptual change. In D. Malament (ed.), *Essays in the History and Philosophy of Science and Mathematics*, 129–166. Lasalle, IL: Open Court.

Norton, J. (2011). *Analogy*. unpublished draft, University of Pittsburg.

Sherrington, D., & Kirkpatrick, S. (1978). Infinite-ranged models of Spin-Glasses. *Physical Review B*, 17(11), 4384–4403.

Yang, G., Lai, C. S. W., Cichon, J., Ma, L., Li, W., & Gan., W. (2014). Sleep promotes branch-specific formation of dendritic spines after learning. *Science*, 344(6188), 1173–1178.

# 7    Biological robustness

## Design, organization, and mechanisms

*Maria Serban and Sara Green*

## Introduction

Systems biology makes extensive use of formal tools from engineering sciences and applied mathematics to explain dynamic features of living systems (Stelling et al. 2004; Wagner 2005; Alon 2007). The success of these epistemic practices has sparked philosophical debates concerning the ability of the mechanistic framework to account for the abstract features of the resulting models and explanations (Brigandt et al. 2018). Many arguments have centered on the adequacy of the mechanistic model of explanation to accommodate strategies drawing on network science and control theory to account for certain types of biological phenomena (e.g., Huneman 2010; Woodward 2013; Skillings 2015; Craver 2016; Chirimuuta 2017; Halina 2018). Focusing on how systems biologists model the phenomena of biological robustness, we aim to clarify the explanatory role of design principles identified via control theoretic analysis.[1]

To illustrate our account, we shall refer to the modelling efforts involved in the investigation of the robustness of bacterial chemotaxis. This case study has been discussed by several philosophers of science without reaching a consensus on whether it entails a causal-mechanistic or a noncausal model of explanation (e.g., Braillard 2010; Woodward 2013; Matthiessen 2017; Green and Jones 2016). This chapter aims to clarify this debate through a more detailed analysis of the mathematical modelling involved in what we describe as an engineering approach to bacterial chemotaxis. Our account offers an alternative to the dichotomy between interpretations of design principles as schemas for more detailed mechanistic explanations (Matthiessen 2017) and noncausal explanations citing *necessary* conditions for the target phenomena (Braillard 2010).

The notion of "design principles" discussed in this paper does not refer to an evolutionary "design" process. Instead, it relies on a "thin" notion of design, referring to "generalizable patterns of organization which play a role-functional part in present-day biological systems" (Green et al. 2015, 16). Modelling efforts in systems biology often describe the dynamics of target systems in terms of abstract patterns of organization that the system implements or realizes, such as negative feedback control or bistable switching. An important feature of design principles is that they are represented abstractly with the help of formal mathematical tools.

The application of design principles in the explanatory projects of systems biology could be broadly analyzed in terms of Pincock's recent account of "abstract explanation" in science (Pincock 2015). However, our analysis of a control theoretic model of biological robustness will show that the abstract character of design principles is not sufficient to capture the distinctive character of the explanation associated with this type of model. While the abstractness of design principles allows for generalizations of biological mechanisms and for the detection of organizational patterns across different biological systems (Levy and Bechtel 2013), we claim that their explanatory value should be analyzed as a separate epistemic contribution. We argue that engineering models of biological robustness draw their explanatory force from a formal mathematical model that can be causally interpreted. The mathematical model offers a derivational warrant for the formal representation of a certain type of design principle instantiated in the target biological system. Design principles have explanatory power in virtue of representing the relational causal *structures* upon which the robustness of a class of biological systems or properties depends. In short, we show how design principles can be discovered via control theoretic engineering analyses and used in "structural-causal explanations" of biological robustness.

The strategy is this. Section 2 explains why robustness is a joint topic of interest to both engineering and biology and introduces the case of bacterial chemotaxis. Section 3 summarizes the modelling efforts developed to understand protein network robustness in the case of bacterial chemotaxis. Here we draw a contrast between a model that interprets robustness as a result of a fine-tuning process and two related models showing that a property called robust perfect adaptation of the *E coli* chemotaxis network depends on certain structural features of this network. Section 4 argues that the design principle introduced by the engineering model – namely, the principle of integral feedback control – figures in a "structural-causal" explanation of the robustness of the bacterium's adaptive chemotactic behavior. Section 5 compares these engineering-based explanations to noncausal design explanations and mechanistic explanations offered in systems biology. In section 6, we summarize our conclusions and clarify how structural-causal explanations differ from canonical mechanistic explanations that represent both causal and constitutive factors responsible for some target phenomenon.

## Modelling robustness – from engineering to biology

Robustness is defined as the maintenance of functional stability in the face of external or internal perturbations (Kitano 2004).[2] Stability, homeostasis, and canalization are forms of biological robustness that have been intensively studied, but they do not exhaust the wide range of robust biological capacities and properties (e.g., Slepchenko and Terasaki 2004; Barkai and Shilo 2007). Alongside studies that focus on different types of biological robustness, there is currently a growing interest in identifying similarities and differences in the ways in which biological and engineered systems achieve robust properties. Since robustness is an important capacity in both engineered and biological systems, an interesting question is whether robustness in both kinds of systems relies on similar design principles (Stelling et al. 2004).

Here we focus on how modelling and conceptual tools from engineering sciences impact the explanation and understanding of cases of biological robustness. Our case study revolves around the theoretical and experimental efforts involved in understanding chemotactic behavior in *E coli*. We start with a brief presentation of the phenomenon of bacterial chemotaxis, narrowing our focus to the feature of robust perfect/exact adaptation. We then contrast a fine-tuned model of the cell's adaptive behavior to two related but distinct models – what we term a "dynamical" and an "engineering" model of the robust perfect adaptation of *E coli's* chemotactic movement. By clarifying the relations between these three models, we aim to pin down the role of design principles in understanding the robustness of a specific type of cellular behavior.

### Bacterial chemotaxis and robust perfect adaptation

Chemotaxis is the process that enables motile bacteria to approach beneficial chemical environments and escape hostile ones. *E coli* swims via movement of external flagella that alternate between two kinds of motions called runs and tumbles (see Figure 7.1). The bacterium executes a "run" by rotating its flagellar motor counterclockwise (CCW). This aligns all of its flagella into a bundle, resulting in a straight line movement for cca. 1 sec. For a "tumble," the bacterium rotates its flagellar motor clockwise (CW). During a tumble, the bundle breaks and the asynchronized flagella produce stationary changes of direction (cca. 0.1 sec). Thus, *E coli* is randomly reoriented after each tumble.

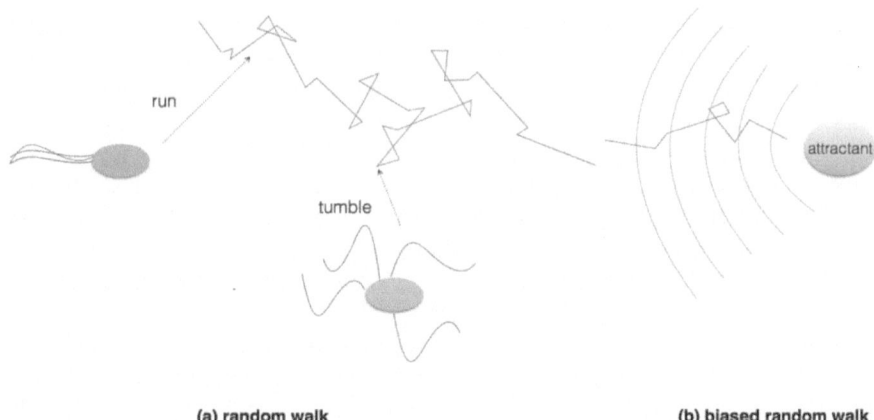

**(a) random walk**         **(b) biased random walk**

*Figure 7.1* Motile behavior of *E coli*.

Source: Adapted from Eisenbach 2001.

Note: This figure can be accessed in color via the eBook version of the book and eResources at www. routledge.com/9780815380788.

The motion of the bacterium in a uniform external environment resembles a random walk since it does not have any ability to control or select its direction of motion. Its straight runs are subject to Brownian motion. In the presence of a chemical attractant, the bacterium moves in the direction of the attractant. Its motile behavior involves less frequent tumbles, which results in longer runs and thus gradual motion toward the attractant (the opposite behavior is observed for repellents such as metal ions or leucine). Due to their small size, most bacteria are unable to sense chemical gradients along the length of their body. Instead, their motion is determined by a "temporal gradient" ($dX/dt$) that captures temporal changes in the chemical concentration of nutrients or toxins in their environment.

*E coli* displays a remarkable sensory adaptation that involves a return to prestimulus levels and receptor sensitivity despite the continuous presence of attractants or repellents. In the presence of a chemical attractant mixed uniformly into the environment at a constant concentration, *E coli's* chemotactic behavior is *perfectly adaptive*. This means that after a period of decreased tumbling frequency (called adaptation time, $\tau$), *E coli's* tumbling frequency ($f$) increases toward and returns to the *exact* same frequency ($f_0$) that it had prior to detecting the attractant in the cell's environment. Remarkably, the receptor system regains its sensitivity and responsiveness to new concentration changes. The *functional stability* of this characteristic adaptive behavior is the *robustness* property that has been of high interest to biologists and engineers alike and which we shall analyze further.

### The chemotactic network

As outlined in the previous section, the ability of *E coli* to migrate in a particular direction depends on the control of the direction of rotation of its flagella. This control is in turn the result of a series of intracellular signals transmitted through a network of interacting proteins and receptors that sense concentration changes. The biomolecular network for *E coli* chemotaxis is currently one of the best understood signal transduction systems. It comprises transmembrane receptors and the adaptor protein CheW, a two-component signaling system (proteins CheA and CheY) whose output controls the motor's tumbling frequency (i.e., the probability that the cell will tumble). Moreover, it involves a protein controlling the output of the signaling system (CheZ) and two proteins, CheB and CheR, that modify CheA's activity via the methylation state of the receptor complex.

The *perfect* adaptation response in *E coli* is the result of interactions between two biochemical pathways that are schematically presented in Figure 7.2. When an environmental attractant attaches to a receptor, the receptor lowers the activity of the CheW-CheA complex. Less activity from this complex reduces the rate of CheY phosphorylation, which results in less CheY-P diffusing to the flagella. CheY-P induces clockwise rotation of the flagellar motor, so decreased levels of CheY-P result in less frequent tumbling (or longer straight runs). After the new attractant has been detected by receptors, the lower activity of the CheW-CheA complex also induces less CheB activity, which means reducing the rate at which methyl groups are removed from the CheW-CheA complex. This, together with

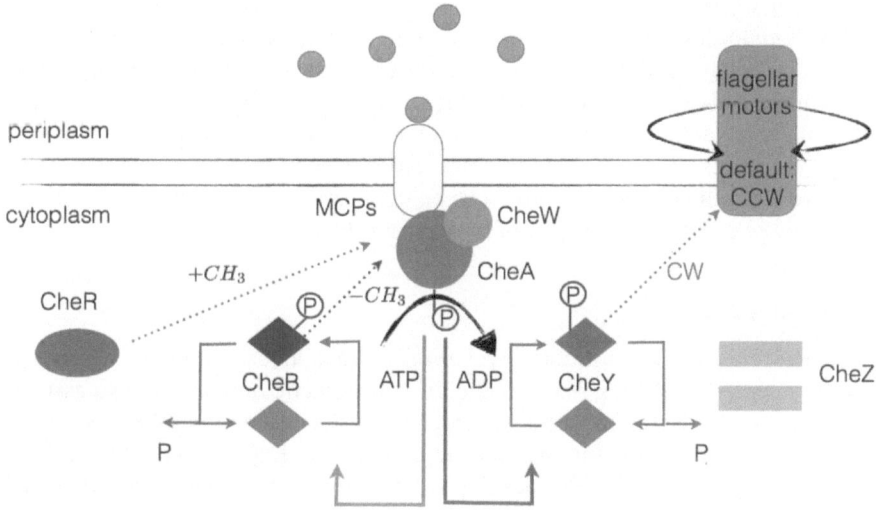

*Figure 7.2* Chemotactic network of *E coli*.

Source: Adapted from Registry of Standard Biological Parts.

Note: This figure can be accessed in color via the eBook version of the book and eResources at www. routledge.com/9780815380788.

the continuous methylation of the CheR receptor, leads to the increased methylation of the CheW-CheA complex. However, more methylation of this complex implies more overall activity. This in turn leads to more phosphorylation of CheY, which will diffuse to the flagellar motor, increasing the clockwise motor rotation and raising the tumbling frequency.

Not only is the chemotactic behavior of *E coli* perfectly adaptive, but the same molecular mechanism (the phosphorylation pathway whose activity is under the control of the methylation pathway) also leads to the chemotactic network returning to the same steady-state tumbling frequency value ($f_0$) irrespective of the concentration value of the protein CheR. That is, the perfect adaptation of the chemotactic behavior is *robust* across a wide range of CheR concentrations. This capacity is commonly referred to as "robust perfect adaptation" (RPA), and in the following section we shall discuss efforts to model this capacity mathematically.

## Modelling the robustness of perfect adaptation

The three models we discuss next capture the efforts of elucidating and explicating *why* the perfect adaptation of *E coli* chemotaxis is robust to CheR protein concentrations ($R$) in the presence of a well-distributed chemical attractant. The first model sketched later on aims to determine whether the perfect adaptation of bacterial chemotaxis is sensitive to the concentrations of the specific proteins that constitute *E coli's* chemotactic network. As we will briefly show, this *fine-tuned* model entails the wrong predictions concerning the behavior of the bacterium. So while it might initially seem the most obvious choice for an explanatory model of the target behavior, the fine-tuned model gets things wrong (cf. Alon 2007). The dynamical and engineering models presented next show that the perfect adaptation of chemotactic behavior is a robust phenomenon, i.e., it remains stable under large variations of chemotactic proteins. After introducing the main elements of the two modelling strategies, we focus on delineating the role that design principles play in explaining this case of biological robustness.

### A fine-tuned model

The fine-tuned modelling strategy relies on a direct, albeit simplified, description of the interactions taking place in *E coli's* chemotactic network. It also assumes that the exact adaptation response depends on the realization of a precise balance between the different chemotactic parameters. For instance, Albert Goldbeter et al. (Knox et al. 1986) developed a fine-tuned model in which adaptation is explained in terms of receptor modification, which acts as a direct response to changed external conditions. According to this model, only methylated receptor molecules trigger activity in a sensory system that is proportional to a weighted combination of the fractions of molecules that are in each of the four possible states. Modelling the dynamics of the receptors' methylation requires tracking the activities of the methylating enzyme CheR and demethylating enzyme CheB. Further assuming that CheR works at saturation with rate $V_R$ while CheB works with Michaelis-Menten kinetics, the model requires very precise tweaking of the concentration rates of the chemotactic proteins to obtain exact adaptation to step increases in stimuli.

In other words, exact adaptation in this model depends on a strict relation between chemotactic proteins so that changing the value of the parameters leads to loss of the perfect adaptation of chemotactic behavior. For example, contrary to what has been observed in the actual behavior of *E coli*, changing the value of CheR concentration by a factor of 20% entails a loss of the precise adaptation response (cf. Alon 2007, 145–146).

Limitations of the fine-tuned model to capture the actual adaptation response observed in bacterial chemotaxis led to the development of an alternative dynamical model. Barkai and Leibler's (1997) modelling strategy aimed to capture the stability of *E coli's* adaptive behavior in the face of wide variations of chemotactic parameters. For this purpose, the researchers abandoned the idea that the adaptive response is the result of a fine-tuning internal mechanism and focused on explaining the robustness (or functional stability) of this cellular behavior.

### A dynamical model of robustness

The first step of the dynamical modelling strategy is to specify when and how chemotactic proteins (CheW-CheA, CheR, CheB, CheY, and CheZ) affect each other. Using the same simplifying assumptions as the fine-tuned model, Barkai and Leibler (1997) built a very simple network with a single receptor species $E$ (standing for the complex MCPs+CheA+CheW), which could be in either one of two states: "active" or "inactive." The active state is characterized by increased CheA activity, which phosphorylates CheY, therefore inducing tumbling. The output of the model is the overall activity of the complex, calculated as a weighted average of all the individual forms of the receptor complex and their activity probabilities (i.e., the average number of receptors in the active state). This quantity is functionally dependent on the CheY-P level and on the kinetic rates of CheY dynamics. The activity probabilities are determined by the input value and the methylation level of the complex. The receptor complex can exist in either attracted/bound or unattracted/free form, and it can be successively methylated. The rates of change of all possibilities resulting from combining the states described above entail the 14 ordinary differential equations (ODEs) system proposed by Barkai and Leibler (1997).

The second step of Barkai and Leibler's modelling consists of writing the relevant equations and implementing them in a computer simulation program. Running the simulations yields a series of predictions concerning the adaptation of the network in the presence of a well-distributed chemical attractant for CheR concentrations varying over several orders of magnitude. That is, the output of the simulation model ($f$) is checked against variations in other network parameters (in particular CheR concentrations). This "sensitivity analysis" shows that the ratio $f/f_0 \to 1$ despite changes in relevant network protein concentrations. This in turn establishes the *robustness* of perfect adaptation since it shows that adaptation does not depend on fine-tuning the values of the various network proteins.

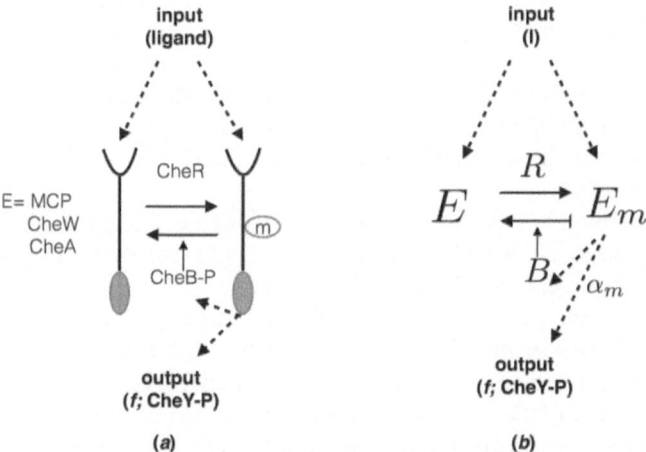

*Figure 7.3*  Representation of the simplified network used in Barkai and Leibler's (1997) model, where $E$ represents the receptor complex in its activated state (non-methylated and methylated), $f$ is the tumbling frequency, and $R$ and $B$ are the concentrations of the methylation proteins CheR and CheB.

Alon et al. (1999) tested experimentally the predictions of Barkai and Leibler's (1997) dynamical model. The experimental manipulation involved controlling the changing concentrations of each of the chemotaxis proteins (which are the target disturbances for the robustness of perfect adaptation). Using genetically engineered *E coli* strains, Alon et al. first deleted the gene for CheR from the chromosome and then introduced into the cell a copy of the gene under control of an inducible promoter. Addition of the inducer led to higher CheR concentration in the cells. The *E coli* population response to a saturating step of attractant was monitored using videomicroscopy. Subsequent runs of the experiment involved changes in the expression levels of all the other chemotaxis proteins. These experimental manipulations confirmed the predictions of the simulations done as part of Barkai and Leibler's modelling strategy, showing the functional stability of the adaptation response to varying CheR concentrations.

## *An engineering model*

Further research showed that there is a formal analogy between *E coli's* chemotactic network robust perfect adaptation and the capacity of some engineered systems to maintain some desired function despite significant external perturbations (e.g., how a car's automatic pilot maintains a constant speed). The model developed by Doyle's group (and published in Yi et al. 2000) identifies a structural similarity between the robustness of the two types of systems, captured by the "integral feedback control feedback principle" (IFC), which is a principle well known from control theory.

Control theory is a mathematical framework that describes the behavior of continuously operating dynamic systems typically in engineered processes and machines. Applied to engineering disciplines, the framework offers algorithms and methods for controlling such systems under specific constraints, such as to avoid delay and overshots and to maintain stability of the control action. Doyle's group relied on these mathematical techniques to reduce the set of ODEs that characterize the main processes in the chemotactic network to an equation that describes formally the action of a specific type of controller used in several engineering systems.

As illustrated on Figure 7.4, IFC establishes that the integral of the difference (or error) between the actual output and the desired (steady-state) output is fed back into the system, thus controlling its activity.

In the following, we explain how this analogy has been established and how it can be used to explore the robustness of some biological properties, like perfect adaptation. The starting point of the engineering model (Yi et al. 2000) was Barkai and Leibler's set of ODEs that was reduced by a series of analytic techniques to a single equation, which corresponds in the control theoretic framework to one of the formulas for computing IFC.[3] Yi et al. (2000) argued that the chemotactic network realizes the type of feedback – abstractly represented by the IFC principle – via the methylation state of the receptor. They also claimed that in both types of systems (biological and engineered), IFC is a *necessary and sufficient* condition

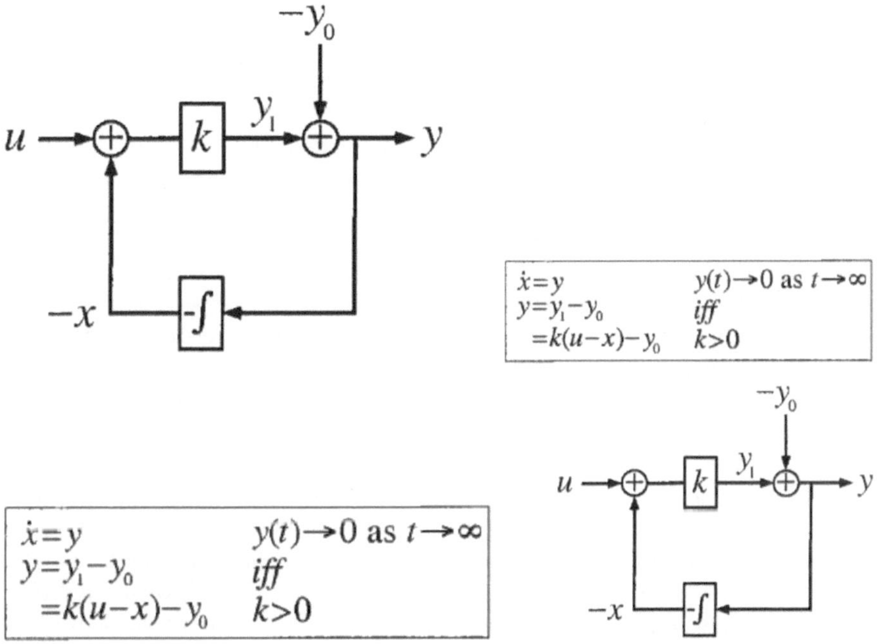
$-y_0$
$u$
$k$
$y_1$
$y$
$-x$
$-\int$
$\dot{x}=y$
$y=y_1-y_0$
$=k(u-x)-y_0$
$y(t)\to 0$ as $t\to\infty$
iff
$k>0$

*Figure 7.4* A block diagram illustrating the IFC principle, which ensures that responses to an input u (here, chemoattractant concentration) are normalized to provide an output y that calibrates the difference between the actual output ($y_1$) and the desired output ($y_0$: steady-state receptor activity) – modelled as the integral of the system error on x.

Source: Yi, T.M., Huang, Y., Simon, M.I., & Doyle, J. (2000) Robust perfect adaptation in bacterial chemotaxis through integral feedback control. Proc Natl Acad Sci USA 97:4649–4653. Copyright (2000) National Academy of Sciences, USA.

for obtaining robust control of some target steady-state property (Yi et al. 2000, 4652). We return to this stronger claim below.

The conclusion that the robustness of perfect adaptation should be characterized and understood in terms of the IFC principle is supported by the mathematical equivalence derived between the initial set of ODEs and the IFC characteristic equation. This equivalence guarantees that there is a formal analogy between the internal organization of engineered and biological systems, which in turn helps explain why both systems achieve robust properties in virtue of instantiating a specific type of organization described by the IFC design principle.

The engineering model also guided the researchers in their exploration of the empirical assumptions underlying Barkai and Leibler's original model. In deriving the IFC equation from the original set of 13 ODEs, Doyle's group argued that only four assumptions are necessary to obtain the perfect adaptation response: (1) CheB demethylates only active receptors; (2) the kinetic rate constants of CheR and CheB are relatively independent of methylation state and ligand occupancy of receptor complex; (3) the activity of unmethylated receptor is negligible relative to methylated receptor forms; and (4) the concentration of bound CheR is independent of ligand level. These assumptions need to be in place to obtain the precise adaptation output either via the simulations or in solving the control engineering equation. Derivation from these (e.g., through mathematical manipulation of the parameters) will not give the desired output. Therefore, Yi et al. argue that variations in the robustness of perfect adaptation behavior can be accounted for in terms of departures from these assumptions.

In our analysis so far we have insisted on the continuity between the dynamic and the control theoretic characterization of the robustness of perfect adaptation. However, it can be argued that the control theoretic model is relatively independent of the specific dynamic assumptions of Barkai and Leibler's model. On the one hand, the formula for IFC can be derived from a different set of ODEs that would correctly approximate the dynamics of the target network. On the other hand, the precise IFC formula depends on the complexity of the system. In other words, there is no unique formula for IFC. The integral of the error between the actual output and the desired output can be fed back into the system in a direct way or via a series of nested feedback loops. Implementing different types of nested feedback loops will yield different formulas for IFC. However, the IFC is sufficiently abstract and general to apply to a variety of cases in both biology and engineering. Therefore, changing some of the empirical or mechanistic assumptions about the target chemotactic network might still lead to a dynamical description that can be reduced to an IFC formula, understood as a more general design principle. As discussed in the following section, this has some important implications for the explanatory characteristics of the model.

## Structural-causal explanations

A first striking feature of the account associated with the control theoretic model is that it characterizes the robustness of perfect adaptation in terms of the *abstract* category of a design principle (the IFC principle). Following the terminology used by the scientists themselves (e.g., Alon 2007), we will sometimes refer to these explanations as "design explanations," although they differ from another class of contrastive functional explanations that have been discussed in the philosophy of biology under the same name (Wouters 2007; van Eck and Mennes 2016, 2018). Section 5 discusses the relevant differences, but for now we stress that our aim is to clarify the sense in which design principles from engineering function as explanatory abstractions[4] in accounts of biological robustness. For this purpose, we can rely on a thin notion of a design principle (Green et al. 2015; Levy and Bechtel 2013) that refers to typically formal representations of ways in which complex systems can be internally organized or relationally linked.

As shown in section 3, characterizing the robustness of the perfect adaptation of bacterial chemotaxis in terms of the principle of integral feedback control was the result of a modelling process that involved "leaving out" various details about

*E coli's* chemotactic network. The result of this process of abstraction is a mathematical formula that characterizes a certain type of internal organization or a pattern of organization that can be implemented either in an engineering or a biological circuit. This representation of the pattern of organization implemented by the chemotactic network during perfect adaptation is *more abstract* than the representation proposed in Barkai and Leibler's dynamical model of the same target phenomenon. By making this comparison explicit, we want to highlight the fact that the abstractness of the design principle used in the engineering-based explanation of biological robustness is a relative feature of the representation being used: one representation or model is more abstract than the other (Levy, 2018). Further, we argue that the more abstract model, despite containing less causal details, is explanatory because it conveys the *right* type of (i.e., relevant, sufficient) information concerning the explanatory questions, i.e., about the generic factors on which the perfect adaptation response depends.

The abstract representation identifies that the target property (that of robust perfect adaptation) depends on a certain form of internal organization of the system (as illustrated in Figure 7.4). To capture the explanatory work that engineering design principles can do in contexts such as that of bacterial chemotaxis research we can repurpose one recent philosophical account of *abstract explanation* in science. Pincock (2015) claims:

> We have an explanation when we have found (1) a classification of systems using (2) a more abstract entity that is (3) appropriately linked to the phenomenon being explained. Whenever an explanation has these three features I will say that we have an *abstract explanation*.
>
> (Pincock 2015, 867, *m.e.*)

Pincock's (2015) account of abstract explanation dovetails nicely with our discussion of the key features of the engineering-based account of robust perfect adaptation. We showed that the engineering model builds upon the *abstract* representation of the chemotactic network in the dynamical model. The formal derivation methods allow researchers to explain the property of robustness in terms of the model incorporating the abstract IFC design principle. The abstract description is "appropriately linked to the phenomenon being explained" because it is derived from the ODEs that describe the behavior of the chemotactic network. In addition, the design principle plays a systematization (or classification) function by grouping together the causal systems (biological and engineered) that exhibit robust behaviors in virtue of the causal structures in which their internal parts (whatever they might be) are embedded (Green and Jones 2016, 355).[5]

Since the systematization function is typically associated with the generality of an explanatory account, it is important to clarify that although engineering explanations using design principles tend to have a wider scope than, for instance, particular mechanistic explanations, their generality is not a mere consequence of their abstract character. That is, we do not take either abstractness or generality to be sufficient to distinguish this class of explanations of biological robustness. As argued by Matthiessen (2017), the mechanistic framework can also accommodate certain types of abstract explanations.[6] Moreover, cataloguing the result of the engineering model as an abstract explanation *tout court* fails to be informative about the role that engineering design principles play in this and other similar cases of biological explanation. For this reason, we need to take our analysis of engineering explanations of biological robustness beyond Pincock's account of abstract explanations. We propose that such explanations should be further qualified as a subtype of abstract explanations in science; namely, as *structural-causal explanations*.

This qualification might strike some as oxymoronic given that current philosophical debates tend to use examples of structural explanations in science to articulate and defend a view of noncausal explanations (e.g., Ladyman and Ross 2009; French 2014; Feline 2017; Glennan 2017). However, there are exceptions to this rule. Some philosophical accounts of structural explanation do not build in a strong contrast between the causal and structural character of an explanation (e.g., Bokulich 2009; Leuridan 2013; Haslanger 2016). To avoid a purely terminological dispute, we proceed next to clarify the two labels that we use in characterizing these explanations of biological robustness.

First, the term "structural" can be directly linked to the thin notion of design introduced in the opening of this section. Thus, we take the notion of "structure" to designate the same thing that "patterns of organization" did in our definition of the thin notion of design; namely, a set of interrelationships holding between the objects that constitute a complex system. Stuart Shapiro offers such a definition of the generic notion of structure:

> A structure is the abstract form of a system, highlighting the interrelationships among the objects, and ignoring any features of them that do not affect how they relate to other objects in the system.
>
> (Shapiro 1997, 73)

We explicitly cite his definition to avoid a potential confusion with another sense of the term "structure," which in biology is often taken to refer to the concrete or material constituents of a biological system. To keep the two senses apart, we will talk about relational structure when the "patterns of organization" meaning is intended. Following this line of thought, abstract explanations such as those that appeal to engineering design principles are structural explanations because they emphasize the link between the relational structure of a system (biological or engineered) and the robustness of some specific functional property or behavior that the system exhibits. An important aspect of the case discussed here is that the model provides a mathematical warrant for the realization of a certain type of organizational pattern and thus for the explanation of the target robustness phenomenon.

Second, we also claim that this class of design explanations has a *causal* character. The formal model articulated in terms of the IFC principle is linked to the original explanation (the robustness of perfect adaptation) via a causal interpretation. However, this causal interpretation does not suffice to justify the explanatory force of the model. The formal mathematical derivation of the IFC equation from the set of ODEs describing the interactions in the chemotactic network is also a key element in securing the relevance of the explanation. Therefore, we claim that there are two features that make the control theoretic model explanatory relative to its target phenomenon (the robustness of perfect adaptation): (1) the mathematical derivation of the IFC equation that provides a mathematical warrant for the robustness of the perfect adaptation, regardless of the many variations of causal details, and (2) the causal interpretation of the variables in the IFC equation that ensures the empirical relevance of the design principle for the case of bacterial chemotaxis. An important aspect to notice is that the causal interpretation of the IFC design principle does not amount to a causal story about how the robust perfect adaptation response is achieved in particular systems. Rather, its role is to guarantee that the variables in the formal model correspond to causal relations in *E coli's* chemotactic network. In addition, the mathematical model can be causally interpreted in the context of other systems (biological or engineered) that exhibit similar robust behaviors. As we will show in section 5, this distinction is important for differentiating structural-causal explanations from mechanistic explanations in biology.

In summary, we maintain that the engineering modelling strategy identifies a particular type of *causal structure* – namely, integral feedback control – upon which the robust adaptive behavior of the bacterium *E coli* depends. The causal structure represented formally via the characteristic equation of IFC explains why the adaptive behavior is robust rather than fine-tuned. That is, it explains why the adaptation occurs despite wide variations in biochemical parameters rather than being sensitive to their precise values.

In more general terms, engineering-based design explanations of robustness work by showing that sometimes the functional stability of some property or behavior of a biological system depends on certain causal structural features of its internal organization. In addition to identifying this type of structural dependence relation, design explanations help to systematize or group together those causal systems (biological or engineered) whose robustness depends on the relational structure in which their components are organized. The fact that such relational structures are typically characterized in formal terms (e.g., by deriving the characteristic equation of the IFC principle, in our case study, from the equations describing the core dynamics of the chemotactic system) is what makes the proposed explanatory account a subtype of abstract explanations in science.[7] By appealing to design principles that represent the causal structures on which the robustness of certain biological properties or behaviors depends, structural-causal explanations are "insensitive" to details about the causal story of the target robust phenomenon. Since they bypass detailed causal descriptions as well as constitutive features of the chemotactic system's adaptive behavior, they differ from canonical causal-mechanistic explanations in biology. We return to this point in the following section.

## Causal but nonmechanistic explanations

The robustness of bacterial chemotaxis has been used before as a case study for philosophical accounts of biological explanation. Here we briefly review two such proposals that use engineering modelling to support noncausal and mechanistic approaches to explanation. We start with Braillard's account of design explanations, highlighting the problems of a noncausal interpretation of the engineering modelling strategy. Then we turn to Matthiessen's mechanistic proposal of viewing design principles as mechanism schemas. We argue that the latter proposal downplays the distinctive character of the modelling strategy outlined in subsection "An engineering model."

### *A noncausal model of design explanations*

The engineering model of robust perfect adaptation discussed in the section "An engineering model" is presented by Braillard (2010) as a clear-cut example of a systems biology strategy yielding noncausal, nonmechanistic explanations of biological robustness. Braillard's approach extends Arno Wouters's (2007) account, according to which "design explanations" are special types of functional explanations that target questions such as, "Why do fish respire with gills rather than lungs? Why do tetrapods respire with lungs rather than gills? Why don't these organisms respire through their skin?" (Wouters 2007). Wouters argues that design explanations differ from both etiological and causal role functional explanations since

> they [design explanations] explain neither by showing that during evolution these structures have been selected nor by showing how the mechanism contributes to some function of the system. Rather they point to synchronic and non-causal functional dependencies between the system's structure and its environment. The aim is to understand why the actual design is better than some alternatives.
>
> (cf. Braillard 2010, 51)

So the explanatory value of descriptions invoking design principles depends, according to Wouters, on the fact that they invoke physical laws or constraints. The explanatory answers to the physiological questions cited connect the organism's need for oxygen, laws of gas diffusion, and the physical properties of the environmental medium, the respiratory surface, and the size of the animal. For instance, it follows from Fick's law of gas diffusion that an organism with a high demand for oxygen needs a large respiratory surface. Design explanations that invoke such general physical constraints or laws are further qualified as types of *abstract contrastive explanations*. They establish that the occurrence of some target feature of a system depends on some physical law or constraint.[8]

Extending Wouters's account from comparative physiology to systems biology, Braillard claims that the explanation of robust perfect adaptation in terms of the

IFC design principle is a *noncausal design explanation*. The core of Braillard's argument for the noncausality of design explanations is that they provide necessary conditions for the functionality of a biological system. He writes, "Design explanations [. . .] show why a given structure or design is necessary or highly preferable in order to perform a function or to have an important property like robustness" (Braillard 2010, 55).

Braillard does not provide a criterion for delimiting causal from noncausal, model-based explanations. It is therefore not clear what blocks the causal interpretation of the explanation associated with the engineering modelling efforts of Doyle's group. Braillard suggests that causal explanations must have temporality built into them. Since design explanations using engineering principles like the IFC do not make explicit the temporal distinction between cause and effect (what precedes what) he concludes that they are noncausal. However, a closer look at the relational structure represented by the IFC equation shows that the temporality of the cause-effect relation is present in the type of feedback corresponding to the IFC principle. According to this design principle, the system is organized so that it uses information about the difference between two states (actual and steady state). This implies that at some time $t_2$, the system's activity (in the case of the bacterial chemotactic system, the level of methylated protein concentrations) depends on something that happened at $t_1$ when the activity level was different from the steady-state ($t_0$) activity levels.

To strengthen the noncausal interpretation, Braillard emphasizes the link between the abstract character and the generality of design explanations. Although abstractness and generality are often thought to go hand in hand, we argue that they are best understood as distinct properties of scientific models. Abstractness concerns details or, more precisely, lack thereof, while generality is about the scope of a model (Levy 2018). So while we agree with Braillard that model-based explanations of robustness that invoke design principles are both abstract and general, we don't think this characterization is incompatible with their causal character. That is, the variables in the formally derived engineering model can be interpreted causally as representing relations in the target chemotactic system.

Moreover, whether the generality of design explanations is a virtue in the context of explaining biological robustness needs to be empirically established via additional scientific modelling and experimentation rather than derived conceptually from the assessment of the abstract character of mathematical models. For this reason, we disagree that the generality of design explanations of biological robustness can be inferred conclusively from the abstract (mathematical) character of the engineering model. What is required is systematic empirical investigation of the scope of applicability of design principles in framing explanations of different instances of biological robustness.

Braillard's interpretation is motivated by Yi et al.'s claim that IFC is a *necessary and sufficient condition* for the robustness of perfect adaptation (and similar biological properties). Such claims are clearly different from mechanistic explanations and possibly instantiate a higher-level description of mechanistic explanations, in the sense that IFC provides "a formal constraint on possible mechanistic models capturing the causal operation of chemotaxis in different systems" (Green 2015, 641). Yet one should be cautious about taking such claims about the scope of models at face value. IFC might be both necessary and sufficient for achieving the robust control of certain steady-state properties in simple systems that operate in a noisy environment. But this is different from showing that the robustness of perfect adaptation *always* depends on the IFC-like organization of biological organisms. Indeed, other researchers commenting on the paper by Yi et al. have cautioned against the idea that RPA necessarily entails the modelled form of integral control (Briat et al. 2016).

Still, the discussion about RPA and the IFC principle shows that an important epistemic aim in this research context is to explore the genetic features of biological (and engineering) systems with the same capacities. The engineering modelling efforts support the search for nested types of control-feedback organization on which the robustness of other biological properties might depend (Stelling et al. 2004; Rao et al. 2004, 2008; Typas and Sourjik 2015). In these contexts, the focus shifts from that of explaining how specific systems (or mechanisms) work to that of investigating the relational structures that can support a specified type of function (Green 2015). While we believe that the noncausal character of Braillard's alternative account of design explanations is disputable, we agree with Braillard that a mechanistic interpretation does not adequately capture the explanatory virtues of the engineering model. We now turn to such an example of a mechanistic interpretation of the same case study.

### Design principles and mechanism schemas

As a foil, we consider a philosophical analysis that argues for the mechanistic character of design explanations of biological robustness. Matthiessen (2017) argues that the abstractness of design explanations does not suffice to demonstrate that they constitute distinctive nonmechanistic types of explanations in biology. His argument is in line with other mechanist philosophers who emphasize the role of abstraction and idealization in the construction and validation of mechanistic explanations (e.g., Abrahamsen and Bechtel 2010; Levy and Bechtel 2013; Craver and Darden 2013: 32–34; Baetu 2015). We agree with this perspective and with much of Matthiessen's criticism of noncausal interpretations (see section 4.1). Yet we take issue with his central contention that the explanatory features of control theoretic analyses of biological robustness can be understood only against the backdrop of a mechanistic framework.

In particular, Matthiessen (2017) argues that the dynamical and engineering models of the robustness of perfect adaptation are explanatory only insofar as they count as mechanism *schemas* for the target phenomenon. More generally, he implies that all explanatory projects pursued in systems biology must satisfy mechanistic standards. To have a mechanistic explanation of a phenomenon implies, minimally, showing how lower-level entities are organized and interact with each other to produce a higher-level target phenomenon (e.g., Machamer, Darden and Craver 2000; Illari and Williamson 2012; Skillings 2015). According to a widespread consensus among philosophers of biology, mechanistic explanations differ from canonical causal explanations in virtue of their citing constitutive or compositional explanatory facts (e.g., Fagan 2015; Glennan 2017). In other words, mechanistic explanations are best understood as hybrid causal-constitutive accounts of target phenomena or regularities, such as the robustness of perfect adaptation. The absence of reference to constitutive relevance factors in the control theoretic explanation of biological robustness is one of the challenges facing a straightforward mechanistic rendition of this type of modelling.

Matthiessen's strategy is to show that abstract representations of engineering design principles can be understood as mechanism *schemas* and that the IFC principle is explanatory only insofar as it is a stage toward a *more complete* mechanistic description of how *E coli* achieves the robust property of perfect adaptation.

A first problem with this strategy is that it points to a contentious issue concerning the standards of mechanistic explanations; namely, the completeness ideal or the idea that the more concrete the explanation the better. This normative ideal implies that nothing short of a complete (highly detailed) description of the microlevel constituents as well as their interactions and modes of organization would satisfy the standards for a mechanistic explanation of a target phenomenon. But Matthiessen seems to take the completeness ideal to be compatible with the explanatory value of engineering design principles interpreted as examples of mechanism schemas. Given the current state of scientific research, and the nature of scientific inquiry, few if any mechanistic descriptions can be expected to meet the completeness standard. So, on primarily pragmatic grounds, mechanism schemas should get the honorific title of being explanatory.

We do not object to the pragmatist defense of the completeness ideal, but rather to the more implicit and problematic claim that the overall (or perhaps ultimate) aim of modelling efforts in fields like systems biology is to elaborate more concrete (or detailed) mechanistic descriptions of target phenomena such as the robustness of perfect adaptation in *E coli* and in other specific bacterial species. This methodological assumption is captured in the statement that biological explanations are *primarily* concerned with particular phenomena ("It is important to retain this sense that explanations in biology are constantly oriented towards particular phenomena." cf. Matthiessen 2017, 18). Granting that more complete (or more concrete) mechanistic explanations are desirable in certain research contexts does not imply that they are always to be preferred. Crucially, the engineering modelling methodology in the case examined here seems to take quite the *opposite* direction – toward enlarging the scope of the applicability of the modelling results and exploring more general underlying design principles (cf. Green 2015).

Hence, one problem with narrowing the focus on particularist explanations in biology is that it fails to do justice to the variety of explanatory practices in modern biology. This challenge is particularly important when examining interdisciplinary practices like systems biology, which may be inspired by explanatory ideals from the physical and engineering sciences. Another serious problem with analyzing the explanatory contributions of control theoretic analyses in terms of mechanistic schemas is that the application of the label "mechanistic" risks becoming a purely stipulative matter (Huneman 2010). This goes against the idea that mechanistic models of explanation aim to capture the standards used in specific scientific practices to evaluate the outcomes of different modelling strategies.

A second related problem is that mechanistic standards require that explanations cite both causal and constitutive factors for some target phenomenon. In addition, as we showed in our analysis of the engineering modelling strategy, constitutive and compositional claims do not figure in control theoretic approaches that analyze robust properties in terms of engineering design principles. This suggests that, even on a minimal construal of the mechanistic standards for explanation, one should avoid hastily classifying design principles as mechanism schemas and pinning their explanatory value to their being part of a continuum of mechanistic descriptions.

One mechanistic interpretation that comes close to our account is formulated by Levy and Bechtel (2013). They grant that design principles can inform about generic features of system organization by specifying how mechanisms with a specific pattern of connectivity will behave. Unlike Matthiessen (2017), they thus argue that sometimes less (causal detail) is more, and that the virtues of design principles cannot be reduced to mechanism schemas. While we agree with this interpretation, we caution against interpreting design principles merely as abstract mechanisms. To show *why* the same behavior occurs in all or most systems in which a specified type of function is realized is a different explanatory project than traditional mechanistic research. Importantly, research on RPA also goes beyond the causal structures of existing systems in exploring potential design principles for artificial systems within systems biology (Briat et al. 2016). We believe that our suggestion of structural-causal explanations better captures the specific explanatory focus of research aiming to identify generic principles and patterns among causally distinct systems.

### *Summary*

In summary, our primary agreement with Braillard (2010) lies in the acceptance of the explanatory irrelevance of many causal details in abstract engineering models. One strength of the engineering model lies in the applicability of the same principle to systems of varying causal details that share characteristic dynamical and structural features. In fact, if explanation is about the identification of dependence relations, then omitting these details is an *epistemic requirement* for selecting and identifying the right factors on which the target robust property depends. On this point we differ from Matthiessen's (2017) interpretation of the same case study. Where we agree with Matthiessen is that the engineering model is not an instance of derivation of necessary and sufficient dependence relations that are insensitive to empirical demonstration of applicability. For instance, an argument that the IFC is also responsible for the robust adaptive behavior in other bacterial species, e.g., *B subtilis*, requires an empirical demonstration showing that this target behavior is similarly insensitive to details about the specific constitution of its chemotactic network.

Finally, we have argued that pointing out the abstract or mathematical character of the engineering model (or of the design principle that figures in the explanations) does not suffice to establish the noncausal character of the explanation associated with it. Many models cite mathematical dependencies between the parameters or variables characterizing a given phenomenon, and some of dependencies represent causal relations, whereas others capture noncausal relations (Glennan 2017, 237). Whether the mathematical content of a model determines the causal or non-causal character of an explanation depends on the kind of dependence relation the mathematical description captures in the world or on the type of question being asked. In the case of the engineering model, the characteristic equation of the IFC principle represents the causal structural features of the organization of *E coli's* chemotactic network. The explanatory force of the model depends *both* on the mathematical warrants provided by the derivation of the IFC characteristic equation and on the causal interpretation of its variables. The *structural-causal* account of design explanations tries to do justice to both sources of the explanatory value of the engineering model.

## Conclusion

Bacterial chemotaxis has been the topic of philosophical as well as scientific debates. The case has received great scientific interest because of the remarkable ability of *E coli* to retain sensitivity to concentration changes, "disregarding" remaining high concentrations of attractant in its environment. Due to this characteristic behavior, the cell can continuously explore and detect novel changes in its environment. In attempting to explain the feature of robust perfect adaptation, a model emphasizing fine-tuning has turned out to have severe limitations by failing to obtain robustness even for small variations in the concentration of the key enzyme CheR. The combined efforts of a dynamic and an engineering model have demonstrated that the robust capacity depends on structural features of the chemotaxis network. We take this case study to raise important philosophical questions about the epistemology of engineering-based approaches to modelling and explaining biological robustness.

In section 3, we showed that the dynamic model (Barkai and Leibler 1997) offers a representation of perfect adaptation in *E coli* chemotaxis that combines a mechanistic schema with a quantitative description of key dynamic features of the chemotactic network. The corresponding set of ODEs was the basis for a simulation study that investigated whether the tumbling frequency variable ($f$) returns to the same steady-state value ($f_0$) despite internal and external changes. If, after a certain adaptation time ($\tau$), the ratio p= $f/f_0=1$, then the adaptation of the bacterium's movement to its environment is said to be *precise*. The simulation-based sensitivity analysis established that changes in the chemotactic protein concentrations leave this ratio invariant.

The strategy adopted by Doyle's group involves the mathematical derivation of the IFC characteristic equation from the original set of ODEs. Since the IFC is a design principle implemented in engineered systems that are built to exhibit a robust behavior, a reverse formal analogy with biological systems suggests that the robustness of perfect adaptation in the *E coli* case *depends* on the fact that its chemotactic network implements the IFC principle. That is, the difference between the current and steady-state output of the system is fed back into the system via the methylation pathway, and this leads to the robustness of perfect adaptation. Accordingly, Yi et al. (2000) argue for a broader inference to the effect that similar types of organization principles can explain the robustness of other cellular behaviors.

Research on the robustness of bacterial chemotaxis thus holds several interesting lessons for understanding the contributions of engineering approaches to biology. In our view, the diversity of scientific practice calls for philosophy of science to accommodate various models of scientific explanations, ranging from explanations of particular events to explanations of general types of behaviors or events observed in a wider variety of systems.

Our aim has been to clarify the *distinctive* character of the explanatory strategy associated with engineering-based modelling. The control theoretic strategy applied by Doyle's group identified a specific design principle or abstract pattern of organization (the integral feedback control principle) as the main factor on which the robustness of perfect adaptation depends. We labelled the result of this modelling strategy a "structural-causal explanation" because the main explanatory category is an abstract representation of a causal relational structure or pattern of organization.

To clarify the explanatory features of the model, we have found reference to abstraction and generality insufficient. Similar examples of design principles have been interpreted as abstract models within a mechanistic framework (Levy and Bechtel 2013). While we generally agree with this account, we believe that interpreting design principles merely as abstract mechanisms risks downgrading epistemic virtues that Levy and Bechtel themselves highlight. Design principles show *why* the same behavior occurs in causally different systems that share a particular organizational structure. While this research question can inform mechanistic research, we regard this as an explanatory project distinct from the mechanistic aim of specifying how certain capacities are realized through interactions between specified causal entities.

We believe that our suggestion of structural-causal explanations better captures the virtues of the engineering model. More specifically, we have argued that the engineering model explains the robustness of perfect adaptation in bacterial chemotaxis by representing a set of interrelationships within the given system. This interpretation highlights relational structures of a system (biological or engineered) that can be described in a formal framework. Our proposal of a structural-causal account thus highlights the empirical connection to the experimental manipulations of the temporal concentration changes in the chemotactic network while also highlighting the fact that the engineering model increases understanding by identifying and ignoring the causal details that are not necessary for describing the dynamics as a result of the relational structure of the system.

# Notes

1 Notably, our focus on "biological robustness" differs from philosophical discussions of "robustness analysis" as a comparative method that takes several models of the same target phenomenon and seeks to determine how reliable or good knowledge of the target is (e.g., Wimsatt 2007; Weisberg 2006, Eronen 2015, Jones 2018). In this chapter, we are concerned with robustness as a property of biological systems and how robustness is modelled and explained in biology and engineering (for a similar focus, see, e.g., Wagner 2005; Huneman 2010, 2017, 2018; Levy 2017).

2 This working definition implies that robustness is a relative, not an absolute, property. No system is robust with respect to all its behaviors (or properties) and under any kind of change. The system as a whole can have both robust and fragile properties or behaviors. In the biological as well as in the engineering sciences, the notions of robustness and functionality are tightly connected because what is taken to be robust is either some typical activity or behavior performed by the system or some desired system characteristic.

3 The steps of the mathematical derivation are shown in the supplementary material of Yi et al. 2001. The derivation focuses on the activity of the receptor complex and applies the principle of mass action to the ODEs in the original Barkai and Leibler model.

4 Similar points about the explanatory value of abstract representations or categories have been made in Jansson and Saatsi (2016); Levy (forthcoming).

5 "Design principles categorize seemingly different biological processes into types through a demonstration of a general principle that they all instantiate, and this categorization makes many causal details or differences between biological processes irrelevant for understanding certain system behaviors. Importantly, unification in this context does not mean reduction of mechanistic explanations to principles. Rather unification is reached through higher-level reflections on types of system organization" (Green and Jones 2016, 355).

6 See also (Levy and Bechtel 2013).

7 By emphasizing the structural elements of design explanations, our account has features in common with "explanations by constraints," a model of scientific explanation recently defended by Huneman (2017, 2018), Green and Jones (2016), and Lange (2017). This family of accounts describes explanations by constraints as "noncausal" and emphasizes the use of mathematical or formal tools in delineating the explanatory answers to the target questions. We have argued that design principles that figure in some structural explanations of robustness can be interpreted causally despite their mathematical representational format. However, this difference does not rule out the possibility of there being noncausal explanations of other types of robustness (cf. Huneman 2017, 2018).

8 Discussing Wouters's account, van Eck and Mennes (2016) characterize the structure of design explanations as follows: "Design explanations address the following type of contrastive why-question: 'why does organism o have trait t rather than trait t?' Or somewhat more fine-grained: 'why does item i have characteristic c rather than c?'"

# References

Alon, U. (2007). *An Introduction to Systems Biology: Design Principles of Biological Circuits*. Boca Raton, FL: Chapman & Hall/CRC Press.

Alon, U., Surrette, M. G., Barkai, N., & Leibler, S. (1999). Robustness in bacterial chemotaxis. *Nature*, 397, 168–171.

Baetu, T. (2015). The completeness of mechanistic explanations. *Philosophy of Science*, 82(5), 775–786.

Barkai, N., & Leibler, S. (1997). Robustness in simple biochemical networks. *Nature*, 387, 913–917.

Barkai, N., & Shilo, B-Z. (2007). Variability and robustness in biomolecular systems. *Molecular Cell: Perspective*, 28, 755–760.

Bechtel, W., & Abrahamsen, A. (2010). Dynamic mechanistic explanation: Computational modeling of circadian rhythms as an exemplar for cognitive science. *Studies in History and Philosophy of Science Part A*, 41(3), 321–333.

Bokulich, A. (2009). How scientific models can explain. *Synthese*, 180(1), 33–45.

Braillard, P. A. (2010). Systems biology and the mechanistic framework. *History and Philosophy of the Life Sciences*, 32(1), 43–62.

Briat, C., Gupta, A., & Khammash, M. (2016). Antithetic integral feedback ensures robust perfect adaptation in noisy biomolecular networks. *Cell Systems*, 2(1), 15–26.

Brigandt, I., Green, S., & O'Malley, M. (2018). Systems biology and mechanistic explanation. In S. Glennan & P. McKay Illari (eds.), *The Routledge Handbook of Mechanisms and Mechanical Philosophy*, 362–374. New York: Routledge.

Chirimuuta, M. (2017). Explanation in computational neuroscience: Causal and non-causal. *The British Journal for the Philosophy of Science*, 69(3), 849–880.

Craver, C. (2016). The explanatory power of network models. *Philosophy of Science*, 83(5), 698–709.

Craver, C., & Darden, L. (2013). *In Search of Mechanisms: Discovery Across the Life Sciences*. University of Chicago Press.

Eisenbach, M. (2001). Bacterial chemotaxis. In *Encyclopedia of Life Sciences*. John Wiley & Sons.

Eronen, M. (2015). Robustness and reality. *Synthese*, 192(12), 3961–3977.

Fagan, M. B. (2015). Collaborative explanation and biological mechanisms. *Studies in History and Philosophy of Science Part A*, 52, 67–78.

Feline, L. (2017). Mechanisms meet structural explanation. *Synthese*.

French, S. (2014). *The Structure of the World: Metaphysics and Representation*. Oxford: Oxford University Press.

Glennan, S. (2017). *The New Mechanical Philosophy*. Oxford University Press.

Green, S. (2015). Revisiting generality in biology: Systems biology and the quest for design principles. *Biology and Philosophy*, 30(5), 629–652.

Green, S., & Jones, N. (2016). Constraint-based reasoning for search and explanation: Strategies for understanding variation and patterns in biology. *Dialectica*, 70(3), 343–374.

Green, S., Levy, A., & Bechtel, W. (2015). Design sans adaptation. *European Journal for Philosophy of Science*, 5, 15–29.

Halina, M. (2018). Mechanistic explanation and its limits. In S. Glennan & P. McKay Illari (eds.), *The Routledge Handbook of Mechanisms and Mechanical Philosophy*, 213–225. New York: Routledge.

Haslanger, S. (2016). What is a structural explanation? *Philosophical Studies*, 173(1), 113–130.

Huneman, P. (2010). Topological explanations and robustness in the biological sciences. *Synthese*, 177(2), 213–245.

Huneman, P. (2017). Outlines of a theory of structural explanations. *Philosophical Studies*, 175(3), 665–702.

Huneman, P. (2018). Diversifying the picture of explanations in biological sciences: Ways of combining topology with mechanisms. *Synthese*, 195(1), 115–146.

Illari, P., & Williamson, J. (2012). What is a mechanism? Thinking about mechanisms across the sciences. *European Journal for Philosophy of Science*, 2(1), 119–135.

Jansson, L., & Saatsi, J. (2016). Explanatory abstractions. *British Journal for the Philosophy of Science*, 70(3), 817–844.

Jones, N. (2018). Inference to the more robust explanation. *British Journal for the Philosophy of Science*, 69(1), 75–102.

Kitano, H. (2004). Biological robustness. *Nature Reviews: Genetics*, 5(11), 826–837.

Knox, B. E. et al. (1986). A molecular mechanism for sensory adaptation based on ligand-induced receptor modification. *Proceedings of the National Academy of Sciences*, 83(8), 2345–2349.

Ladyman, J., & Ross, D. (2009). *Every Thing Must Go: Metaphysics Naturalized*. Oxford University Press.

Lange, M. (2017). *Because Without Cause: Non-Causal Explanations in Science and Mathematics*. Oxford University Press.

Leuridan, B. (2013). The structure of scientific theories, explanation, and unification: A causal-structural account. *British Journal for the Philosophy of Science*, 65(4), 717–771.

Levy, A. (2017). Causal order and kinds of robustness. In E. Lamm, S. Gissis, & A. Shavit (eds.), *Landscapes of Collectivity in the Life Sciences*. MIT Press.

Levy, A. (2018). Idealization and abstraction: Refining the distinction. *Synthese* (2018). https://doi.org/10.1007/s11229-018-1721-z.

Levy, A., & Bechtel, W. (2013). Abstraction and the organization of mechanisms. *Philosophy of Science*, 80(2), 241–261.

Machamer, P., Darden, L., & Craver, C. (2000). Thinking about mechanisms. *Philosophy of Science*, 67, 1–25.

Matthiessen, D. (2017). Mechanistic explanation in systems biology: Cellular networks. *The British Journal for the Philosophy of Science*, 68(1), 1–25.

Pincock, C. (2015). Abstract explanations in science. *British Journal for the Philosophy of Science*, 66, 857–882.

Rao, C. V., Glekas, G. D., & Ordal, G. W. (2008). The three adaptation systems of bacillus subtilis chemotaxis. *Trends in Microbiology*, 16(10), 480–487.

Rao, C. V., Kirby, J. R., & Arkin, A. P. (2004). Design and diversity in bacterial chemotaxis: A comparative study in Escherichia coli and bacillus subtilis. *PLoS Biology*, 2(2): e49.

Shapiro, S. (1997). *Philosophy of Mathematics: Structure and Ontology*. Oxford University Press.

Skillings, D. (2015). Mechanistic explanations of biological processes. *Philosophy of Science*, 82(5), 1139–1151.

Slepchenko, B. M., & Terasaki, M. (2004). Bioswitches: What makes them robust? *Current Opinioni in Genetics and Development*, 14(4), 428–434.

Stelling, J., Sauer, U., Szallasi, Z., Doyle, F. J., & Doyle, J. (2004). Robustness of cellular functions. *Cell Review*, 118, 657–685.

Typas, A., & Sourjik, V. (2015). Bacterial protein networks: Properties and functions. *Nature Reviews: Microbiology*, 13, 559–572.

van Eck, D., & Mennes, J. (2016). Design explanation and idealization. *Erkenntnis*, 81(5), 1051–1071.

van Eck, D., & Mennes, J. (2018). Mechanism discovery and design explanation: Where role function meets biological advantage function. *Journal for General Philosophy of Science/Zeitschrift für Allgemeine Wissenschaftstheorie*, 49(3), 413–434.

Wagner, A. (2005). Distributed robustness versus redundancy as causes of mutational robustness. *Bioessays*, 27(2), 176–188.

Weisberg, M. (2006). Robustness analysis. *Philosophy of Science*, 73(5), 730–742.

Wimsatt, W. C. (2007). *Re-engineering Philosophy for Limited Beings: Piecewise Approximations of Reality*. Harvard University Press.

Woodward, J. (2013). Mechanistic explanation: Its scope and limits. *Aristotelian Society Supplementary*, 87(1), 39–65.

Wouters, A. (2007). Design explanation: Determining the constraints on what can be alive. *Erkenntnis*, 67(1), 65–80.

Yi, T-M., Huang, Y., Simon, M. I., & Doyle, J. (2000). Robust perfect adaptation in bacterial chemotaxis through integral feedback control. *Proceedings of the National Academy of Sciences USA*, 97(9), 4649–4653.

# Part 3

# Societal issues

# 8 The machine analogy in bioethics

*Andreas T. Christiansen*

## Introduction

> *This will not appear at all strange to those who know how wide a range of different automata or moving machines the skill of man can make using only very few parts, in comparison to the great number of bones, muscles, nerves, arteries, veins, and all the other parts which are in the body of every animal. For they will consider this body as a machine which, having been made by the hand of God, is incomparably better ordered and has in itself more amazing movements than any that can be created by men.*
>
> Descartes, *Discourse on Method*, part V

Descartes famously believed that animals (and human bodies) were analogous to machines, only designed and built by an infinitely greater engineer. And, (in)famously, the analogy between animals and machines was used by the followers of Cartesian mechanistic philosophy[1] to justify animal experimentation that utterly disregarded the suffering of the animals (e.g., vivisection). Thus, understanding biological entities as machine-like may very well have important practical consequences. Mindful, perhaps, of the Cartesian example, many philosophers and lay debaters are critical of contemporary uses of organism-machine analogies. In this chapter, I critically evaluate such "antimachine" views. I will focus mainly on the debate surrounding synthetic biology, as this is the domain in which the machine analogy, as well as objections to it, is most prevalent. However, as will become clear, many of the substantive ethical objections made by antimachine views have counterparts in other areas of bioethics, including in debates over human enhancement, geoengineering, and agricultural biotechnology.

## Two machine analogies

There are (at least) two machine analogies at play in synthetic biology. One draws an analogy between the *structure and workings* of organisms and those of machines. Call this the "scientific analogy." The purpose of the scientific analogy is to generate a better understanding of organisms or subparts thereof (such as DNA) by using concepts related to machines, such as modules, programs, or part-whole relationships, to explain the organization and functioning of organisms.

As far as I can tell, most of the other chapters in this book are concerned almost exclusively with this version of the organism-machine analogy.

However, novel biotechnologies, and synthetic biology in particular, have effected the emergence of a different machine-organism analogy. One important strand of synthetic biology research aims at introducing concepts of rational design taken from nonbiological engineering domains into biotechnology (Endy 2005). This includes the creation of an inventory of standardized biological parts, known as BioBricks, and efforts to make the design and creation of new biological systems less complex and divisible into smaller tasks. The aim is to increase the efficiency and ease of biological engineering and ultimately to construct living systems that have useful functions, such as chemicals production, medicine delivery, or carbon sequestration. Practitioners and proponents, as well as critics, of synthetic biology frequently talk of "biological machines" in this context. This second analogy, which we might call the "technological analogy," has a dual purpose. First, it serves to highlight the potentials of using biological systems as machines – i.e., of letting organisms do or make things that we want them to do or make. Second, it introduces concepts originating in the design and fabrication of machines, such as standardization and modularity, into the design and fabrication of biotechnological products. Thus, it changes the way in which biotechnology is *done*, making the engineering of living systems more like the engineering of machines.

The division within synthetic biology between scientific and technological is not confined to analogies – it is a division between different aims and projects that, although often related, are distinct. In the political and ethical debate, it is the technological aspect of synthetic biology that has attracted attention. For example, in a report prepared by three scientific committees under the European Commission, synthetic biology is defined as "the application of science, technology and engineering to facilitate and accelerate the manufacture and/or modification of genetic materials in living organisms" (European Commission 2014). Likewise, it is the technological analogy that is the target of the majority of (though not all) antimachine views. And the two analogies are importantly different from an ethical point of view. In particular, the technological analogy is much more closely related to practice than the scientific one. It describes a certain set of actions with respect to organisms; namely, designing and using them analogously to how one would design and use a machine. It is therefore, in a sense, a *normative* analogy. The scientific analogy, by contrast, invites a certain way of viewing organisms that enables understanding; it is a primarily *descriptive* analogy. However, as we shall see, those who voice objections to the scientific analogy supply some link between this way of (passively) viewing organisms and some problematic way of acting toward organisms. It is thus not the scientific analogy per se that is objectionable (according to these critics) but the fact that it invites or entails ways of acting that are objectionable.

## Antimachine views

Generally speaking,[2] analogies are based on some noticed similarity between a "source" and a "target." In our case, the source system is a type of machine (or a part of a machine) and the target system is an organism (or a part of an organism) – for example, the source might be a computer's software and the target DNA. The similarity in question can take several forms, from a simple shared feature to complex, systemic features such as causal patterns. Based on the observed similarity, the analogy user then seeks to generate knowledge or hypotheses about the target based on already existing knowledge about the source. Roughly, the scientific analogy seeks to use knowledge and concepts that pertain to the source system of machines to generate new insights and hypotheses about how the target systems of organisms work. The technological analogy uses knowledge about engineering principles pertaining to machines to generate ideas about how to better design and build biological systems; it also uses knowledge about how machines can be used for human purposes to stimulate the development and use of organisms for human purposes.

Antimachine views focus especially on two distinct aspects of what machines are and how we understand them: (1) machines and their functioning are understood in *mechanistic* terms, where system-level effects are the consequence of the complex interactions of subsystems, modules, or parts; and (2) machines are *artifacts* designed by human beings with human purposes in mind. While the former pertains to both the scientific and technological analogies, the latter pertains exclusively to the technological analogy (since merely understanding organisms by analogy with machines does not make organisms into artifacts). As I will argue later, the main weakness of antimachine views is that they push analogies further than is necessary and further than their original users' intentions.

## Mechanistic understanding

Several critics of synthetic biology point out that it relies on a mechanistic understanding of life. That is, it is a presupposition of synthetic biology that organisms' properties and behaviors can be explained in terms of the interaction of their parts or modules (which are similarly explainable in terms of the interactions of their subparts). This presupposition is inherent in the uses of both the scientific analogy and the technological analogy. But what is the supposed problem with this view? Critics suggest two candidates. First, the mechanistic view is descriptively inadequate or invites neglect of uncertainties and lack of predictability, leading to the creation or lack of prevention of adverse consequences of our interactions with organisms. Second, the mechanistic view destroys the possibility of granting moral standing or patienthood to living beings. I will discuss each of these in turn.

### Descriptive inadequacy and hubris

According to this version of the antimachine view, machine analogies entail a mechanistic understanding of organisms that, in turn, is descriptively inadequate or hubristic. Consequently, using a machine analogy creates risks of adverse effects since it results in poor predictions of organisms' behavior, invites excessive confidence in our ability to make such predictions, or both. An example of this view can be found in the writings of Joachim Boldt:

> Finally, the products of synthetic biology are conceived of as complex chemical machines and thus referred to as entities that can be fully understood in terms of internal regularities. . . . Given the machine model of life, future behavior of an organism appears to be fully determined by the initial conditions of its molecular makeup. Consequently, in order to guard against unexpected behavior, one will first of all define as precisely as possible the initial states and make sure that the actual makeup of the novel organism corresponds to the design as accurately as possible. Now, if it is correct . . . that the development of an organism must be understood in terms of an unpredictable process of action and reaction of the organism in its environment, controlling the initial molecular state of an organism will not suffice to reduce risks. Rather, retrospective knowledge about an organism's interaction with its environment appears to be a more reliable basis for predictions, which implies that risk assessment of novel organisms, if lacking a close natural relative, will necessarily appear to be precarious.
>
> (Boldt 2013a, 398, 400–401)

Let us look a little more closely at Boldt's claims. First, he claims that "given the machine model of life, future behavior of an organism appears to be fully determined by the initial conditions of its molecular makeup." I presume this is true, at least insofar as it is a statement of a form of basic physicalism or ontological reductionism – i.e., of the view that, ultimately, everything is determined by the basic laws of physics.

Next, he suggests that it is a consequence of this view that "in order to guard against unexpected behavior one will first of all define as precisely as possible the initial states and make sure that the actual makeup of the novel organism corresponds to the design as accurately as possible." This claim goes further than the previous claim: it is one thing to affirm that organisms' behavior is *determined* by the initial conditions of its molecular makeup and quite another to claim that we can rely on knowledge and control of such conditions to *predict and control* behavior. The former is a purely metaphysical claim about the ultimate causes of behavior, the latter an epistemological claim about our knowledge of behavior.

Finally, Boldt suggests that assuming "the development of an organism must be understood in terms of an unpredictable process of action and reaction of the organism in its environment, controlling the initial molecular state of an organism will not suffice to reduce risks." In addition, we need "retrospective knowledge about an organism's interaction with its environment" to do so. Here Boldt takes yet another step, suggesting that a mechanistic or machine-analogical understanding of organisms precludes an alternative understanding that emphasizes (quasi-)purposeful interaction of the organism with its environments. The machine analogy thereby also precludes an approach to prediction (and, by extension, risk assessment and risk management) that emphasizes observation of how organisms behave in the environment – or at least it pushes us away from such an approach.

With respect to paradigmatic machines, perhaps all three of these claims hold: (1) machines' workings are determined by the initial configuration of their internal makeup, (2) we can predict and control effects with reasonable accuracy based on knowledge and control of such initial configuration, and (3) we do not resort to explanations of machine "behavior" in terms of (quasi-)purposeful responses to the environment. But, for one thing, the two latter claims are not necessarily true of highly complex machines or machines programmed to learn from interactions with the environment. Furthermore, engineers do not assume that they can adequately predict how a newly designed device will function merely by knowing in detail how it is designed. At any rate, the use of a machine analogy does not *imply* either the second or the third claim. At the very least, the machine analogy is compatible with acknowledgment of complexity in organization – preventing predictability from initial conditions – and with acknowledgment of the necessity of taking a view of organisms as purposeful beings when attempting to understand behavior. And, arguably, the machine analogy may even counteract a tendency to assume that we can predict behavior from the initial molecular makeup alone because it stresses exactly such complexities, feedback mechanisms, and interactions between the different parts of the organism as well as between organism and environment (Holm 2015; O'Malley 2009, 2011).

Boldt's risk-based antimachine view thus depends on stretching the "standard" machine analogy, which merely uses machine concepts to understand organisms. He suggests that using a machine analogy must also mean (1) using machine-analogous methods for predicting and controlling organisms' behavior and (2) *not* using modes of understanding for organisms that we would not use for machines. This extension is certainly not necessary. It is perfectly possible for people, and especially professional biologists, to use the standard machine analogy without thereby believing (1) or (2). Perhaps there is some risk that non-biologists will be susceptible to such a conceptual slide, but that is, at best, a contingent matter that needs empirical support.

### Moral standing

A second type of objection to the mechanistic (or "reductionist") understanding of organisms inherent in the machine analogy holds that such understanding precludes granting appropriate moral standing to organisms. "Moral standing" denotes a property of an entity that confers on it the ability to be wronged (or benefitted) morally. Thus, an entity's possessing moral standing implies some restrictions regarding how others may act relative to – or how they may treat – that entity. The overall idea is that living beings have moral standing in virtue of being purposeful beings, and that the mechanistic or reductionist presuppositions of the machine analogy therefore undermine the moral standing of living beings. A few examples:

> More precisely, it is the reductionist concept of life underlying the project of designing synthetic organisms that is questionable from an ethical perspective. The BioBricks program, for instance, is typically framed along a Cartesian view. Living systems are made of functional parts that can be assembled into a machine like a watch. This mechanistic concept of life supports a project of exploitation. . . . [T]he Cartesian theory of animals-machines conveniently supported the project of using them as machines in the service of men. In stark contrast, Kant's comparison between mechanical machines (watch) and organic bodies (tree) emphasized the distinctive value of organized bodies. Unlike mechanical machines which have a "motive power," organisms have a "formative power." They self-reproduce, they maintain themselves. . . . In this view living entities are an end in themselves and should not be used as machine-tools for heteronomous ends.
>
> (Bensaude-Vincent 2013, 374)

> This may threaten the view that life is special. At least since Aristotle, there has been a tradition that sees life as something more than merely physical. This provides the basis for belief in the interconnectedness of all living things, and the sense that living things are, in some important way, more than organized matter. The special status of living things and the value that we ascribe to life may therefore be undermined by reductionism.
>
> (Cho et al. 1999, 2089–2090)

> Finally, from the perspective of synthetic biology the assumption of an inherent value of life must become untenable. This does not only hold for microorganisms but for any organism that synthetic biology may turn its attention to. As long as synthetic biology adheres to the epistemological and ontological assumptions of fabrication and thus construes itself strictly along the lines of engineering principles, it will not be able to understand its objects as inherently valuable, the fact notwithstanding that the higher an organism one turns to, the more will the assumption of value suggest itself.
>
> (Boldt 2013a, 401)

A number of things are worth noting. First, the quoted authors do not always distinguish clearly between the scientific analogy and the technological analogy. Both Bernadette Bensaude-Vincent and Boldt oscillate between stressing the reductionist view of organisms (the scientific analogy) and the engineered, designed, or artifactual nature of synthetic-biological organisms (the technological analogy). And, indeed, the view that the technological machine analogy serves to undermine organisms' moral standing is a common one, which I shall return to later.

Second, the authors presuppose that living beings do in fact have moral standing *qua* living beings, and that this standing is tied to a view of living things as fundamentally different than nonliving beings due to their being (quasi-)purposeful. As Boldt puts it immediately after the quoted passage, they possess (albeit at a low level) "the ability to act according to what one regards as good." In other words, they assume a biocentric view based on purposefulness or "having a good" (Goodpaster 1978). While they are right to presuppose that purposefulness is the main property in virtue of which biocentrists accord moral standing to living beings, the authors' intimations that biocentrism is a generally accepted view is false. Quite the contrary – outside the narrow confines of environmental ethics, biocentrism is very much a minority view. Most people do not exhibit any constraints on how they act toward bacteria or yeast, nor are they troubled to learn of the atrocities they regularly visit upon these creatures.

Third, the authors quoted seem to assume that not only the moral standing of microorganisms is in peril but also the moral standing of all living beings, including animals (and perhaps even human beings). But this is likely an error. Perhaps the machine analogy precludes a view of living beings *qua* living as purposeful and thus undermines the argument for moral standing of living beings *qua* living based on purposefulness. But from the fact that living beings do not have moral standing *qua* living it does not follow that no living beings have moral standing. And several other grounds for moral standing have been suggested, most importantly sentience and sophisticated agency. The former ensures moral standing to sentient animals, the latter to human beings and perhaps some cognitively sophisticated animals such as apes or ravens. Perhaps the worry is that ordinary folks, lacking philosophical sophistication, would be prone to draw the mistaken inference from "not all living beings have moral standing" to "no living being has moral standing." But this is, at best, highly uncharitable and probably false (Douglas et al. 2013, 692).

The reductionism-undermines-moral-standing version of the antimachine view thus has certain external problems. But it also has internal problems – problems that arise even if we assume that it is problematic if organisms are not seen as purposeful beings. And, again, the problem lies in overstretching the machine analogy. It is true that the machine analogy requires mechanistic explanation and understanding of organisms, but it need not be reductionist. That is, it need not *deny* the purposefulness perspective. In fact, the mechanistic and the purposeful views of organisms are complementary rather than mutually exclusive. This is the crucial difference between the current use of machine analogies and the Cartesians' analogy between animals and automata. In the latter case, the analogy's core tenet was a *denial* of the basis for moral standing; namely, sentience. In the former case, the users of the analogy merely affirm similarities between organisms and machines but not similarity with respect to those properties of machines that mean they do not have moral standing.

Perhaps even affirming bona fide reductionism would not undermine a biocentric view of the moral standing of organisms. Consider the related debate concerning the relationship between determinism and moral responsibility.

Seeing people as morally responsible is to accord them with a kind of moral standing: the burden of responsibility goes with such benefits as rights, freedom, and respect (Strawson 1962). This debate is, I would suggest, simply a debate between the compatibility of purposefulness-based and mechanistic perspectives on human action – the former being necessary for responsibility, the latter being more or less identical with determinism (as it relates to human action). Here we can see how debates in other parts of bioethics turn on the same issues that are involved in antimachine views. Debates concerning cognitive – in particular moral – enhancement of human beings frequently touch upon the question of whether seeing human beings' cognitive and moral abilities as something that *can* be manipulated somehow undermines the possibility of also seeing human beings as rational and responsible agents. Similarly, the use of, for example, neuroscientific or social psychological knowledge to predict and control behavior – e.g., in the context of preventing crimes – is controversial in part because it implies that behavior is so predictable.

This similarity with the responsibility/determinism debates suggests that the conflict between mechanistic and purposefulness-based understandings is a real one and thus may support the antimachine views currently under consideration. Nevertheless, it seems to me that two lessons can be drawn from the responsibility/determinism debates, both of which suggest that the antimachine view is not persuasive. First, while the debate surrounding responsibility and determinism is obviously ongoing, compatibilism is at least a live option. According to compatibilism, assigning moral responsibility to agents is compatible with there being ultimately deterministic explanations of their actions. This is partly because, as Strawson argued, it is hard, if not impossible, to relinquish the view of humans as purposeful.

Thus, even if one believes that it is problematic, or in some sense false, to deny that organisms are purposeful (albeit in a different sense than humans), one can accept the use of machine analogies as long as they merely affirm a mechanistic perspective without thereby denying the purposefulness perspective. This would yield a compatibilist biocentrism – i.e., the view that seeing organisms as purposeful is compatible with there being ultimately mechanistic explanations of their behavior. And such a view seems a more attractive option than the hard-nosed view that also denies the truth of the mechanist perspective. For (and this is the second lesson) it is a question of fact whether the mechanistic perspective is correct. In the realm of human action, compatibilism is attractive because it seems equally difficult to deny determinism as it is to deny the purposefulness perspective. It is no less difficult to deny mechanism, in the most basic sense, where organisms are concerned. In other words, it is not the use of the machine analogy but the general thrust of science that threatens the viability of hard-nosed versions of biocentrism – ones that turn on *denying* mechanism rather than on *affirming* purposefulness. Consequently, I suggest that biocentrists should develop a compatibilist version of their view and, in so doing, dispense with their aversion to machine analogies.

## Artifacts

Apart from being understood and explained in mechanistic terms, machines have another property that antimachine views react against: they are artifacts. For our purposes, artifacts have two important features: (1) they are *designed and fabricated* by human beings with human purposes in mind and (2) they are *used* for the furthering of those purposes. It is thus the technological machine analogy that is the target for this set of antimachine views. As a consequence, these antimachine views may seem to be more surely grounded than the ones above since the technological analogy is apt only where organisms are or have been designed, fabricated, and used for human purposes. Thus, the technological analogy does not merely see organisms *as* artifacts, it also advocates that organisms be in fact transformed into artifacts, or presupposes that organisms have already been so transformed. Nevertheless, I shall suggest that the antimachine views here have exactly the same type of weakness as the ones above: they assume that the scope of the machine analogy is wider than it need be and wider than intended.

As mentioned, one version of the artifact-based antimachine view holds that it is seeing organisms as artifacts, rather than (or as well as) understanding them mechanistically, that undermines moral standing. Apart from this, there are two other variants of the artifact-based antimachine view. One holds that artifactualness undermines the *intrinsic value* of organisms. The other holds that seeing organisms as artifacts expresses an objectionable attitude toward other living beings or nature as such, such as domination or a desire for complete control. I will deal with each of these in turn.

### *Moral standing redux*

Let me again start by quoting some examples of the view I am discussing (some of which belong to the same authors as those quoted earlier):

> All of this [machinist] vocabulary identifies organisms with artifacts, an identification that, given the connection between 'life' and 'value,' may in the (very) long run lead to a weakening of society's respect for higher forms of life that are usually regarded as worthy of protection.
>
> (Boldt and Müller 2008, 388)

> From a biocentric perspective, is it right to conceive living entities as machines performing functions for humans, to reduce them to chassis that can be functionalized in our service? Whatever their degree of artificiality, synthetic organisms are living beings. As such they have an intrinsic value. For instance, the OncoMouse designed in a Harvard laboratory for use as a research tool, which became a patentable invention and a commercial entity, [is] a mouse that suffers so that we may live. This transgenic and human designed object nevertheless raises compassion and solidarity.
>
> (Bensaude-Vincent 2013, 374)

> For positions arguing for an intrinsic value of organisms, their instrumentalization needs to be well justified in order to be morally acceptable. The purpose-oriented perception of entities in synthetic biology avoids this discussion by defining entities as machines, as soon as the aspect of their function for human purposes predominates. No one denies that it is justified to instrumentalize machines, as it is part of their definition that they perform a particular task for human beings.
>
> (Deplazes and Huppenbauer 2009, 62)

There are two aspects to this antimachine view. First, there is a worry that seeing organisms as machines will erode an attitude that is prevalent, or at least can be found, among ordinary people. Thus, the worry is that seeing organisms as artifacts will lead to mistreatment of organisms, perhaps "in the (very) long run," including animals and human beings. Second, some might worry that, if an organism is an artifact, it thereby *in fact* loses its moral standing. Both of these views have problems.

With respect to the first, it is (as noted earlier) doubtful whether many people do in fact hold that living beings have moral standing *qua* living, and thus that there is anything that can be eroded. It is even more doubtful that any erosion that might occur would spread to the moral standing of "higher forms of life," including human beings.

With respect to the second, the problem is that it is hard to see that anything troublesome remains. After all, if an organism's being an artifact entails its not having moral standing, then we cannot be doing anything wrong to that organism. The view that it is somehow wrong to create an entity that we cannot wrong has a certain air of paradox, at least if the wrong done is supposed to consist of a wronging of that very entity. The confusion possibly stems from a conflation of moral standing and intrinsic value since creating an entity lacking intrinsic value, or turning one type of entity into another that lacks such value, may be problematic (to be discussed next). Furthermore, the majority view seems to be that artificiality does not matter for the moral standing of organisms because moral standing is grounded in intrinsic properties, such as sentience or "having a good," which are not affected by the artificiality of the organism (see, for example, Attfield 2012; Baertschi 2012b; Deplazes-Zemp 2012; Huesken 2014; Sandler 2012).

Let me look more closely at one attempt, this time by Christopher Preston (2013a, 122–123), at arguing for an antimachine view of this type. Preston gives two arguments for why artifact-organisms have no, or only a diminished, moral standing.[3] First, he argues that the fact that these entities are *organisms* is only a "concomitant" or "incidental" attribute of them. Their artifactual nature, especially the purpose for which the organism was designed, by contrast, is a necessary feature of the entity. For example, an organism designed to capture and bind carbon in the atmosphere is, on Preston's view, necessarily a carbon-capturing device and only incidentally an organism. Therefore, "the sense in which synthetic bacteria are tools serving a particular purpose takes precedence over the sense in which they are autonomous organisms" (Preston 2013a, 123).

Preston's second argument is that an artifact-organism has a "bifurcated tel-eology." On the one hand, it has an internal, "organismal" teleology like natu-ral organisms have. On the other hand, it has an external, "artifactual" teleology imposed by the designer and defined by her goals and purposes in designing the organism. Moral standing is conferred only on the basis of organismal teleol-ogy: entities that possess only artifactual teleology, such as thermostats, are typi-cally not thought to have moral standing. Since artifactual teleology is primary in artifact-organisms, their moral standing is at least diminished relative to natural organisms. As support for this claim, Preston argues that the fact that people are willing to treat domesticated animals much worse than wild animals suggests that they take domesticated animals (i.e., partially artificial organisms) to have lower moral standing than wild animals (i.e., purely natural organisms).

The main problem, for both of Preston's arguments, is that it stretches the machine analogy. The technological machine analogy stresses that (some) organ-isms should be looked at, from a practical point of view, as machines – both with respect to *how* we design and build them and in the sense that we *can* design and use them for various purposes. It does not, however, suggest that they should not *also* be looked upon as organisms. Again, this variant of the antimachine view slides from the use of a machine analogy that draws upon *some* similari-ties between organisms and machines to the view that organisms are seen as also having *other* properties of machines or as being entirely like machines – in our case, in particular, that they have the further normative property of machines of lacking moral standing as a matter of fact or in the public eye. There is, it seems to me, no strong reason to suspect that the technological machine analogy should or will become stretched in this way. To see why, consider the two arguments Pres-ton makes, which are effectively arguments to the effect that organism-machines' machine-like normative properties dominate their organism-like normative properties.

With respect to the second of Preston's arguments, it rests on a premise that seems to me highly questionable; namely, that people are willing to treat domestic animals much worse than wild animals. First, the view that the welfare of domes-tic animals is inconsequential seems to me to be very much a minority position today. Although mistreatment of agricultural animals is of course a real problem, it seems that it is mainly a matter of privileging business interests than denying that animal welfare matters. Second, even if some people do think domestic ani-mals' welfare does not matter, they may simply be wrong. Third, how people are willing to treat wild and domestic animals varies with the relationship they have to those animals (Anderson 2012). Some domestic animals are seen as members of the family, to be pampered rather than mistreated. And some wild animals are viewed as enemies (e.g., rats), while others are seen as majestic objects of awe and admiration (e.g., blue whales). Finally, it is doubtful that those who *do* deny the importance of domestic animals' welfare also think that *wild* animals' welfare is important.[4]

With respect to the second argument, an obvious objection is that, in a case where a person performs some function, such as checking tickets in the subway, we may not treat her as a mere tool. That is, we may not act toward her as if her dominant normative properties were those of a function-performing entity, much less as though her humanity-based normative properties were nonexistent. Quite the contrary: the ticket checker's dominant moral properties are those she has as a human being – both in fact and in the majority's opinion. Preston counters that the difference can be explained by the fact that being a ticket checker is incidental to the *very existence* of the person, while it is not incidental to the very existence of a carbon-capturing bacterium. But even in cases where performing a function is not merely incidental to a person's existence, it is not the case, nor do most people believe, that we may treat that person as a mere tool. Consider the case of "savior siblings" – children born to provide biological materials that could save an ill older brother or sister (see Boyle and Savulescu 2001). It is a standard and effective objection to allowing savior siblings that they may end up being treated as mere tools. If the intuition that savior siblings may *not* be so treated were not very solid and widespread, this objection would not have the considerable bite that it does. This suggests that the general consensus position described earlier, according to which intrinsic features of entities ground moral standing, is on the right track.[5]

There is another view that objects to viewing organisms as machines in the technological sense. According to this view, we (sometimes) wrong organisms by designing and creating them. We do so if (when) we design organisms to have properties that are strongly detrimental to their well-being. The "OncoMouse" mentioned by Bensaude-Vincent, which is engineered to develop cancer, would perhaps count as a being that is wronged in this way (see also Attfield 2012, 86). But there are other examples, including ones from the context of more traditional breeding, such as chickens unable to carry their own weight. Even in cases such as this, where the way an organism is designed results in predictable suffering, some might deny that the organism is being wronged. A version of the nonidentity problem (Parfit 1984) arises since the alternative to existing as OncoMice or factory chickens is not existing at all. The debate surrounding nonidentity is much too complicated for me to enter here, but it seems to me plausible to at least object to the creation of organisms that will have lives not worth living. This is not to say that creating such organisms may not be acceptable, all things considered. For example, it may be the case that the benefits to cancer research justify the suffering of OncoMice. Cases such as the OncoMouse and the factory chicken may be instances where the maltreatment of organisms that antimachine views fear will result from use of the machine analogy actually occurs. There is certainly maltreatment of animals in agriculture and laboratories. Whether the use of machine analogies is responsible for it is perhaps more debatable. At any rate, it is possible to use the machine analogy, even in its technological sense, without thereby ignoring the suffering of animals. It is almost certainly false, at least, that the use of the machine analogy associated with synthetic biology has caused maltreatment since most instances of maltreatment occurred before synthetic biology even existed.

## Intrinsic value

The second category of objection concerns the "intrinsic value" of organisms. Intrinsic value, as I understand the term here, is distinct from moral standing, in that it does not depend on whether the entity can be wronged. A great work of art can have intrinsic value – it can be good in itself that it exists – but it does not plausibly have moral standing.

Several authors have argued that *natural* living beings possess a value that *artificial* living beings lack. Perhaps the most thoroughly worked out version of this view is Keekok Lee's (1999). On Lee's definition, an entity is an artifact if, and to the degree that, it "embod[ies] human ends" (Lee 1999, 37). Artificiality is thus a degree term for Lee. The main determinant of the degree of artificiality is the "deepness" of the technology used, which for Lee increases as a technology intervenes at more fundamental levels. The ultimate artifact would be one that was built directly from fundamental matter (i.e., atoms, subatomic particles, strings, or whatever turns out to be at the ground-floor of physical reality). The implication for the domain of biotechnology is that lifeforms that are the product of human interventions can be ranked according to their level of artificiality: Traditional trial-and-error selection produces organisms that are less artificial than those resulting from scientifically informed breeding and hybridization, and these are again less artificial than organisms created through genome-level interventions (Lee 1999, 52). Although Lee does not talk about machines, it is easy to see that organisms created according to the recommendations of the technological machine analogy – using standardized modules, typically at the subcellular level – are highly artificial on Lee's account.

Lee's claim is now that naturally occurring entities have "independent value" – a value that is grounded in the very fact that natural entities are independent of humans and human intentions (Lee 1999, 178). Since more artificiality is tantamount to a higher degree of dependence on human design and human aims, an organism loses intrinsic value as it increases in artificiality. In his alternative but closely related view, Preston (2008) initially rejects Lee's view of what grounds the intrinsic value of natural organisms, suggesting instead that it is the fact that organisms are connected to the historical process of evolution that endows them with such value. Later (2013a, 2013b), Preston accepts that the lack of independence in Lee's sense can at least also diminish the intrinsic value of organisms.

Views similar to Lee's and Preston's are prominent in debates on geoengineering and restoration of nature. For example, the restoration of natural ecosystems has been denounced as "faking nature" (Elliot 1982) and a "big lie" (Katz 1992) because restored ecosystems lack the independence of human intervention that generates intrinsic value. Similarly, Boldt (2013c) and others have argued that artificial organisms do not contribute to valuable biodiversity since only naturally evolved species are intrinsically valuable.

As before, the intrinsic-value-based antimachine view turns on a move from the claim that an organism possesses some machine-like feature to the claim that it thereby lacks or loses another, non-machine-like feature. But here the antimachine view seems to be right since the property of independence does in fact seem to be lost or diminished when organisms or ecosystems are also artifacts. (The claim that artifact-organisms are disconnected from evolution seems false, though.) However, the view that intrinsic value depends on independence can be contested. Some, including Lee, seem to derive this premise from a certain respect for the independent (quasi) point of view of other lifeforms and from a non-anthropocentric account of value. These views seem to me confused. It presumably matters little to these other lifeforms whether they are independent of humans. Similarly, it is strange for a supposedly radically non-anthropocentric view of value to assign supreme importance to an entity's relationship with human beings. More importantly, intuitions concerning the intrinsic value of organisms or ecosystems likely rely on something other than independence, such as their grandeur or complexity or beauty. Although the fact that grandiose, complex, and beautiful things have come into being without humans being responsible surely contributes to their value, the view that any and all natural entities, from individual organisms to ecosystems, have intrinsic value seems to me controversial.

For three reasons, however, the persuasiveness of the intrinsic-value-based antimachine view can be questioned. First, it is perhaps a problem to destroy things of intrinsic value, but it is hard to see how creating things that lack intrinsic value is a problem per se. Lee frequently speaks of new biotechnologies as being "nature replacing," but at least in the literal sense it is not true that they replace natural beings. In synthetic biology, nothing is being replaced at all. In agricultural biotechnology, entities that are highly artificial – namely, modern hybrid crop varieties – are replaced by slightly more artificial beings in the form of GM crops.

Second, artifact-organisms have (one should hope) compensating *instrumental* value. It is thus not the case that we bring something valueless into existence, which might be objectionable or irrational. And neither can it straightforwardly be claimed that we could have brought more value into existence by bringing intrinsically valuable entities into existence; the view that intrinsic value is categorically "better" than instrumental value is false (or at least in need of further defense).

Finally, artifact-organisms may possess intrinsic value that is grounded in something other than their independence. For example, they may embody human ingenuity and creativity comparable to a work of art, an invention, or even a paradigmatic machine like a car. In addition, artifact-organisms (and ecosystems) retain all the properties, other than independence, that contribute to grounding the intrinsic value of their natural counterparts, such as beauty or grandeur. My intuition is that, for example, a majestic but artificial tree or ecosystem would possess intrinsic value partly for the same reasons its natural counterparts do (even if independence also contributes to the value of the latter).

We thus see a variant of the overstretching problem that haunts antimachine views: they move from the claim that artifact-organisms lack one property that natural organisms have, and that contributes to the intrinsic value of the latter, to the claim that artifact-organisms do not have any properties that ground intrinsic value. They thus overlook (1) that even paradigmatic machines may be intrinsically valuable, although for different reasons, and (2) that artifact-organisms (or ecosystems) may retain other value-grounding properties of organisms (or ecosystems), other than independence, that paradigmatic machines do not have.

### *Attitudes*

The final antimachine view that I shall discuss holds that viewing organisms as machines is tantamount to taking up some objectionable attitude, such as the desire to dominate nature or to control everything. The idea is that we view organisms as machines because we want to optimize or improve them and to shape them into things that do what we want them to do. Views of this sort are prominent in other domains of bioethics as well. For example, in the domain of agricultural biotechnology, Henk Verhoog argues that "genetic engineering fits into [an] industrial approach to agriculture, because nature is not seen organically, but as a mechanism and an object of human analysis, control and interference" (Verhoog 2007, 395). Similar views are also prominent in the debate on human enhancement. For example, Michael Sandel argues:

> The deeper danger is that [enhancement] represent[s] a kind of hyperagency – a Promethean aspiration to remake nature, including human nature, to serve our purposes and satisfy our desires. The problem is not the drive to mechanism, but the drive to mastery. And what the drive to mastery misses and may even destroy is an appreciation of the gifted character of human powers and achievements.
>
> (Sandel 2009, 78)

Indeed, antimachine views of this sort are central to conservative thought in bioethics and beyond. Thus, Russell Kirk in *The Conservative Mind* argues that conservatives "are dubious of wholesale alteration. They think society is a spiritual reality, possessing an eternal life but a delicate constitution; it cannot be scrapped and recast as if it were a machine" (Kirk 1994, 8).

While the attitudes-based antimachine view is widespread in several domains, it is also notoriously difficult to get a grasp of what exactly the problem is supposed to be (what, for example, is the nature and value of "the gifted character of human powers and achievements"?). One possibility is that the attitude of dominance *just is* objectionable or undesirable. Lee argues that it amounts to narcissism and egomania and is therefore "either pathological or immature" (Lee 1999, 201). A second possibility is to argue that such an attitude is inconsistent with granting living beings moral standing or intrinsic value. Lee suggests that the attitude of dominance and its attendant striving for control "necessarily involves trampling on the legitimate ethical demands of nonhuman others" (ibid., 202). A third possibility is that the attitude is detrimental to *our own* interests. Lee laments the prospect of a world where "[w]herever one turns, one only sees images of oneself; wherever one shouts, one hears only one's own echoes." In such a world, humanity becomes "lonesome" and "imprisoned within an existential or ontological solipsism of its own making" (ibid., 194). A similar view is expressed by G.A. Cohen (2011, 207), who argues that "the attitude that goes with seeking to shape everything to *our* requirements . . . contradicts our own spiritual requirements" and is "repugnant, and, at the limit, insane."

In each of these cases, the problematic aspect of the attitude of mastery or domination or control is linked crucially to its being "totalitarian": it is an attitude of complete mastery of everything that is *really* problematic. To assume that such an attitude is inherent in the use of the machine analogy is false. Once again, the analogy is stretched, so that the desire to improve or even master *one* kind of organism or to optimize in *one* respect that (let us concede for the sake of argument) is inherent in the technological machine analogy is equated with a desire for complete control over every aspect of everything. The point is not merely that biotechnologists are typically not narcissist megalomaniacs: the attitude-based views do not (or do not only) concern such mental states, but rather what Boldt, borrowing a phrase from Hans Jonas, calls "the basic precepts of one's doing" (Boldt 2013b, 40): the question is whether the attitude of mastery is necessary to make sense of the action in question. As far as I can tell, it is not. It is perfectly possible to make sense of, say, someone wanting to enhance his own or his future children's cognitive abilities without ascribing to him the desire to be able to remake himself entirely or to design every aspect of his child. Likewise, it is possible to understand the synthetic biologist who makes a synthetic yeast that produces vanillin without assuming that she must hold the view that every aspect of nature should exist only if it pleases her and only in the form most useful to humanity.

With respect to synthetic biology, it is doubtful that the use of the technological analogy is expressive of even very rudimentary attitudes of mastery. As Tim Lewens (2013) argues, the engineering approach to life that synthetic biologists undoubtedly take is better explained as a set of pragmatic ways of dealing with the complexity of organisms by simplification than as the expression of a rational-design view of life. Thus, the attitude-based antimachine view, like the others, is guilty of interpreting the machine analogy as having more content than it need have and as making more expansive claims than has been intended.

## Conclusion

I have argued that objections to using analogies between machines and organisms are not generally persuasive. Although there are many variants of this type of objection, they all suffer from the same flaw: they extend the scope of the analogy beyond what is necessary or intended by those using the analogy. Those who object to the scientific version of the machine analogy – whereby viewing organisms in machine-terms contributes to our understanding of how organisms work – mistakenly assume that taking a mechanistic view also means *never* taking a more holistic or purposefulness-driven view of organisms. And those that object to our designing and using organisms as we would design and use machines – i.e., those who object to the technological version of the machine analogy – mistakenly assume that designing and using organisms in this way entails a denial that organisms (or, more broadly, nature) have any value that paradigmatic machines do not have. Thus, I suggest, we should treat antimachine views primarily as *warnings* not to make the mistake of overextending the analogy between organisms and machines.

## Notes

1  Although Descartes may himself have held a more moderate view on the treatment of animals; see Cottingham (1978) and Harrison (1992).
2  This section draws on Holyoak and Thagard 1996.
3  Preston argues that artificial organisms have less "intrinsic value" than natural ones, but he does not clearly distinguish between moral standing and intrinsic value. The reasons described here seem to me to pertain mainly to moral standing. It is, furthermore, not entirely clear whether he is making the argument that moral standing would be eroded because of mistaken reasoning from artifactualness to lack of standing or, rather, than artifactualness in fact undermines moral standing.
4  Some environmental ethicists, in particular J. Baird Callicott (1980), argue that domesticated animals have less value than wild animals. But even Callicott does not defend a differential *moral standing* for wild and domestic animals since he (1) initially (in the 1980 paper) denies moral standing to *individual* wild animals as well and (2) later softens his position considerably on both domestic and wild animals (see Lo 2010).
5  Somewhat ironically, then, the very effectiveness of the mere-tool objection to savior siblings suggests that the bad outcome on whose possibility it relies is not, in fact, very likely to occur, and thus that the objection should not be given very much weight.

## References

Anderson, E. (2012). Animal rights and the values of nonhuman life. In C. R. Sunstein & M. Nussbaum (eds.), *Animal Rights: Current Debates and New Directions*. Oxford: Oxford University Press.

Attfield, R. (2012). Biocentrism and Artificial Life. *Environmental Values*, 21(1), 83–94.

Baertschi, B. (2012b). The moral status of artificial life. *Environmental Values*, 21(1), 5–18.

Bensaude-Vincent, B. (2013). Ethical perspectives on synthetic biology. *Biological Theory*, 8, 368–375.

Boldt, J. (2013a). Life as a technological product: Philosophical and ethical aspects of synthetic biology. *Biological Theory*, 8, 391–401.

Boldt, J. (2013b). Creating life: Synthetic biology and ethics. In G. E. Kaebnick & T. H. Murray (eds.), *Synthetic Biology and Morality: Artificial Life and the Bounds of Nature*. Cambridge, MA: MIT Press.

Boldt, J. (2013c). Do we have a moral obligation to synthesize organisms to increase biodiversity? On kinship, awe and the value of life's diversity. *Bioethics*, 27(8), 411–418.

Boldt, J., & Müller, O. (2008). Newtons of the leaves of grass. *Nature Biotechnology*, 26, 387–389.

Boyle, R. J., & Savulescu, J. (2001). Ethics of using preimplantation genetic diagnosis to select a stem cell donor for an existing person. *British Medical Journal*, 323, 1240–1243.

Callicott, J. B. (1980). Animal liberation: A triangular affair. *Environmental Ethics*, 2, 311–338.

Cho, M., Magnus, D., Caplan, A. C., McGee, D., & the Ethics of Genomics Group. (1999). Ethical considerations in synthesizing a minimal genome. *Science*, 286, 2087, 2089–2090.

Cohen, G. A. (2011). Rescuing conservatism. In R. J. Wallace, R. Kumar, & S. Freeman (eds.), *Reasons and Recognition: Essays on the Philosophy of T.M. Scanlon*. Oxford: Oxford University Press.

Cottingham, J. (1978). A brute to the brutes? Descartes' treatment of animals. *Philosophy*, 53(206), 551–559.

Deplazes, A., & Huppenbauer, M. (2009). Synthetic organisms and living machines: Positioning the products of synthetic biology at the borderline between living and non-living matter. *Systems and Synthetic Biology*, 3, 55–63.

Deplazes-Zemp, A. (2012). The moral impact of synthesising living organisms: Biocentric views on synthetic biology. *Environmental Values*, 21(1), 63–82.

Douglas, T., Powell, R., & Savulescu, J. (2013). Is the creation of artificial life morally significant? *Studies in History and Philosophy of Biological and Biomedical Sciences*, 44, 688–696.

Elliott, R. (1982). Faking nature. *Inquiry*, 25(1), 81–93.

Endy, D. (2005). Foundations for engineering biology. *Nature*, 438, 449–453.

European Commission. (2014, September 23). *Opinion on Synthetic Biology I: Definition.*

Goodpaster, K. E. (1978). On Being Morally Considerable. *The Journal of Philosophy*, 75(6), 208–325.

Harrison, P. (1992). Descartes on animals. *The Philosophical Quarterly*, 42(167), 219–227.

Holm, S. (2015). Is synthetic biology mechanical biology? *History and Philosophy of the Life Sciences*, 37(4), 413–429.

Holyoak, K. J., & Thagard, P. (1996). *Mental Leaps: Analogy in Creative Thought*. Cambridge, MA: MIT Press.

Huesken, S. (2014). Artificial life and ethics. *NanoEthics*, 8, 111–116.

Katz, E. (1992). The big lie: Human restoration of nature. *Research in Philosophy and Technology*, 12, 231–241.

Kirk, R. (1994). *The Conservative Mind*, 7th ed. Washington, DC: Regnery Publishing.

Lee, K. (1999). *The Natural and the Artefactual: The Implications of Deep Science and Deep Technology for Environmental Philosophy*. Lanham, MD: Lexington Books.

Lewens, T. (2013). From *bricolage* to BioBricks™: *Synthetic biology and rational design. Studies in History and Philosophy of Biological and Biomedical Sciences*, 44, 641–648.

Lo, Y. S. (2010). The land ethic and callicott's ethical system (1980–2001): An overview and critique. *Inquiry*, 44(3), 331–358.

O'Malley, M. A. (2009). Making knowledge in synthetic biology: Design meets kludge. *Biological Theory*, 4(4), 378–389.

O'Malley, M. A. (2011). Exploration, iterativity and kludging in synthetic biology. *Comptes Rendue Chimie*, 14(4), 406–412.

Parfit, D. (1984). *Reasons and Persons*. Oxford: Oxford University Press.

Preston, C. J. (2008). Synthetic biology: Drawing a line in Darwin's sand. *Environmental Values*, 17(1), 23–39.

Preston, C. J. (2013a). Synthetic bacteria, natural processes and intrinsic value. In G. E. Kaebnick & T. H. Murray (eds.), *Synthetic Biology and Morality: Artificial Life and the Bounds of Nature*. Cambridge, MA: MIT Press.

Preston, C. J. (2013b). Evolution and the deep past: Intrinsic responses to synthetic biology. In R. L. Sandler (ed.), *Ethics and Emerging Technologies*. Basingstoke: Palgrave Macmillan.

Sandel, M. J. (2009). The case against perfection: What's wrong with designer children, bionic athletes, and genetic engineering. In J. Savulescu & N. Bostrom (eds.), *Human Enhancement*. Oxford: Oxford University Press.

Sandler, R. (2012). Is artefactualness a value-relevant property of living things? *Synthese*, 185, 89–102.

Strawson, P. F. (1962). Freedom and resentment. *Proceedings of the British Academy*, 48, 1–25.

Verhoog, H. (2007). Organic agriculture versus genetic engineering. *NJAS-Wageningen Journal of Life Sciences*, 54(4), 387–400.

# 9 The machine metaphor in science and science communication

*Sune Holm*

## Introduction

Scientists are under increasing pressure to communicate their research and results to lay people. The importance of science communication is evident in at least three contexts. First, there is a growing emphasis on the importance of disseminating research paid for by taxpayers to the public through a variety of media channels. Second, to achieve funding, there is an increasing focus on the need to be able to present research ideas to funding bodies and their board members, who will not always be experts in the field of investigation. Finally, policy makers and regulators rely on input from specialists when deliberating about how to legislate with respect to some domain of research and technology.

Arguably, the single most important and frequently applied means of communicating science to nonspecialists is the use of metaphors. Within the life sciences, which will be the focus of this chapter, descriptions of biological concepts, phenomena, and structures often trade in metaphorical expressions. Thus, to point to just a few widely used metaphors, natural selection is described as a tinkerer or a master engineer, genomes are blueprints for organisms, cells are chemical factories consisting of molecular machines, and brains are computers. Metaphors also play a central role in the communication *within* scientific communities, where they can be seen to drive hypothesis formation and development of new research areas and suggest research strategies.

In this chapter, I will review the role of the machine metaphor in science and science communication. I will focus on the relatively young discipline of synthetic biology, which is in many ways organized around machine metaphors for living systems. For clarity, I will simply use the expression "the machine metaphor" to cover the wide range of metaphors used in the life sciences that juxtapose living systems with man-made mechanical or electronical machines such as cars and computers. I will argue that, in general, despite having a well-established heuristic role within a scientific context, the machine metaphor may be an ill-chosen means of communicating the selfsame scientific research and insights to a lay audience. This is because the communicative use of this heuristically useful metaphor may easily slide into a misleading theoretical identification of organisms with machines. Moreover, I back up the philosophical criticism of the use of the machine metaphor in science communication with empirical evidence showing how strongly metaphors may frame our way of thinking about a topic.

The plan of the chapter is this. Section 2 introduces the interaction view of metaphor developed by Max Black, which I think is the most fruitful way of thinking about metaphors in the context of science. In section 3, I describe the engineering ideals and associated machine metaphors driving synthetic biology. Then, in section 4, I present a two-fold categorization of the uses of metaphors in science, the merits of which I discuss in the context of synthetic biology. I show that the machine metaphor is a valuable, though potentially misleading, heuristic in synthetic biology and that it pervades external communication about synthetic biology. I argue that we should be highly skeptical of using metaphors serving a heuristic function in science as a means of communicating scientific knowledge and methods to a lay audience. In section 5, I support my theoretical claims by pointing to recent empirical research that suggests that metaphorical framing, as understood in the interactionist tradition, will often have a significant influence on people's understanding and reasoning about a topic.

## The interaction view of metaphor

In general, when we use metaphor, we describe one kind of thing using terms ordinarily used to describe some other kind of thing. For example, one might say that "crime is a virus," "life is a rollercoaster," "cells are molecular machines," or "Man is a wolf." Contemporary philosophical discussion of metaphor takes as its point of departure Max Black's interaction account of metaphor (Black 1955, 1993). In his examination of metaphors, Black claimed that, on the traditional view, metaphors highlight similarities between a target and a source. Thus, according to "the substitution view," a metaphorical expression is saying something of a target that could be stated literally by some other sentence describing the target. To use Black's own example, when we say, "The chairman ploughed through the discussion," there is no more to the metaphorical expression than a literal expression stating that "the chairman dealt summarily with objections, or ruthlessly suppressed irrelevance, or something of the sort" (Black 1955, 278). Thus, Black characterizes the substitution view as one according to which "understanding a metaphor is like deciphering a code or unravelling a riddle" (Black 1955, 280). Similarly, if I say, "Life is a rollercoaster," then this is simply a way of pointing to certain qualities that can be attributed to both life and rollercoasters. In this way, a metaphor can also be said to express a comparison.

Black argued that there is more to metaphor than the comparison of qualities between target and source. On the interaction view, metaphors do not merely trade in similarities and analogies between target and source; they also "create" similarities and analogies between target and source. On this account, to say that "Man is a wolf" suggests a way of thinking about men (and wolves) that goes beyond simply pointing out that men and wolves share certain features, such as preying on other animals. He writes:

> A suitable hearer will be led by the wolf-system of implications to construct a corresponding system of implications about the principal subject. But these implications will not be those comprised in the commonplaces normally implied by literal uses of "man." The new implications must be determined by the pattern of implications associated with literal uses of the word "wolf." Any human traits that can without undue strain be talked about in "wolf-language" will be rendered prominent, and any that cannot will be pushed into the background. The wolf-metaphor suppresses some details, emphasises others – in short, *organizes* our view of man.
>
> (Black 1955, 288)

To illustrate his idea, Black imagines that he is given the task of describing a battle in the terminology of chess:

> These latter terms determine a system of implications which will proceed to control my description of the battle. The enforced choice of the chess vocabulary will lead some aspects of the battle to be emphasized, others to be neglected, and all to be organized in a way that would cause much more strain in other modes of description. The chess vocabulary filters and transforms: it not only selects, it brings forward aspects of the battle that might not be seen at all through another medium.
>
> (Black 1955, 288–289)

Black's interaction account of metaphors thus takes metaphors to be more than literally false statements that can be substituted for literally true ones without a loss of content. Metaphors are *constructive*, in that they enable us to reason about the target subject on the basis of its likeness with the source. Once we apply a metaphor, it thus comes to *control, transform*, and *organize* the way we think about the target. On the interaction view, metaphor introduces a framework for thinking about a target domain that is *dynamic* in the sense that similarities between target and source will develop in the course of thinking about the target in light of the source. Moreover, the metaphor is *bidirectional*. When we say that "Man is a wolf," this may just as well shape our way of thinking about wolves as it may shape our thinking about man. Also, given the dynamic nature of the interaction view, metaphors will *not* be substitutable without loss of content.[1]

Reynolds (2018) provides a useful illustration of Black's account of metaphor in relation to the factory metaphor of the cell:

> In saying that the cell is a factory [. . .] we suggest that a whole set of relations among members of the semantic field of factories (inputs, outputs, assembly lines, shipping and receiving, energy and waste byproducts, etc.) are applicable to the conceptual domain of cells. The structure of these borrowed

relations becomes the basis for analogical inferences and hypotheses about the target subject.

(Reynolds 2018, 152)

It is noteworthy that in his example Reynolds emphasizes that the analogy highlights that a set of relations characteristic of the source can be applied to the target and form the basis of two core aspects of the scientific endeavor: inferential reasoning and hypothesis generation.

## The machine metaphor in synthetic biology

I now turn to consider in more detail the role of the machine metaphor in synthetic biology. Descriptions of synthetic biology as a new field in its own right tend to differentiate synthetic biology from traditional biotechnologies such as breeding and recombinant genetic engineering by emphasizing that synthetic biologists focus on applying rational design principles to the construction and redesign of biological systems.[2] The rational design of mechanical or electronic machines proceeds by breaking down the design problem into subtasks that can be worked at independently, constructing devices that solve these subtasks and assembling devices into a modular, hierarchical system that has the capacity to perform the desired function. The vision of rationally designing living systems is the vision of developing the tools and techniques necessary for the systematic and hierarchical construction of living systems using standardized parts with well-defined functions. The realization of the ideal of the rational design of a biological system may thus be described in terms of three steps (Endy 2005):

1   Produce standard biological "off-the-shelf" parts with well-defined functions and predictable interactions.
2   Assemble parts into more complex devices.
3   Assemble devices into even more complex devices until the final system is constructed.

A fourth component of the engineering effort of synthetic biology is the production of a minimal genome comprising only the set of genes necessary for survival and reproduction in a particular environmental setting. A cell with a minimal genome is desirable because it can serve as a "chassis" for building more complex systems by plugging in biological parts and devices (Heinemann and Panke 2006, 2793).

The overall motivation for the focus on rational design and construction of parts and modules is to reduce the complexity of living systems found in nature. Tom Knight points out that "an alternative to understanding complexity is to get rid of it" (Ball 2004). Endy (2005) and Keasling (2005) emphasize that the project of enabling rational design is a means to replace the overwhelming complexity of naturally evolved organisms with transparent and predictable systems to ensure efficient production of much more streamlined, profitable, and secure biotechnology. Reducing complexity will ensure "reliable functional composition," meaning that "it would be nice if we had standard biological parts that could snap together and then behave as expected when we snapped them together" (Endy 2008). Hence the need for introducing standardization, decoupling, and abstraction into biotechnology (Endy 2005; Keasling 2005).[3]

The engineering approach to biology is commonly characterized using the machine analogy. Machines are assembled from standardized components to produce reliable behavior that can be derived from the properties of the parts and their interaction.[4] Andrianantoandro et al. (2006) propose an analogy between computer engineering and synthetic biology. In the computer engineering hierarchy, "every constituent part is embedded in a more complex system that provides its context." The bottom physical layer includes transistors, capacitors, and resistors analogous to the DNA, RNA, proteins, and metabolites in a cell. The parts of the physical layer are combined to produce the devices of the device layer – such as logic gates – which is equivalent to the layer of biochemical reactions and regulatory mechanisms in the cell. At yet a higher level of the computer engineering hierarchy, the module layer, we find integrated electronic circuits, which parallel the idea of biological devices that can be plugged into the host cell to enable it to perform a desired function. To list a couple of other examples, the name of one of the most popular annual events within the synthetic biology community is the International Genetically Engineered Machine Competition (iGEM), which envisages bacteria as machines, and synthetic biologists are widely appealing to the notion of "genetic circuits" in the context of intracellular molecular biology (Yokobayashi et al. 2002). It goes without saying that the notion of a "chassis" genome presented earlier is also an instance of the machine metaphor.

In sum, synthetic biology provides a clear and current example of the use of the machine metaphor to organize a research community's thinking about a kind of system.

## Two functions of metaphors

So far, I have presented the interactionist account of metaphor and the use of the machine metaphor in synthetic biology. To structure the discussion, it will be helpful to employ a rough categorization of the ways in which scientists employ metaphors. For the purposes of this chapter, I will make use of a two-fold characterization according to which metaphors can be said to have a rhetorical or heuristic function.[5]

The rhetorical use of metaphor is, according to Bradie, prevalent in the context of, e.g., pedagogy, communication, and political rhetoric (1999, 161). A classic example of rhetorical use of metaphor is Churchill's famous remark that "[f]rom Stettin in the Baltic to Trieste in the Adriatic an 'iron curtain' has descended across the continent." Churchill made rhetorical use of metaphor to communicate how he thought about the political conditions in Europe. Metaphors are also often used to educate and communicate about complex problems. Economists might say that the housing market is "collapsing," while climate scientists try to explain the mechanism underlying global warming in terms of "the greenhouse effect." In these cases, metaphors are employed to convey scientific or specialist knowledge about phenomena to nonspecialists.

To illustrate the heuristic use of metaphors, Bradie uses the example of thinking about gas molecules as if they are billiard balls in random motion on a billiard table, bouncing off each other and the sides of the table. Using metaphors heuristically is a common way for scientists to come up with hypotheses about a domain of investigation on the basis of their understanding of another domain. In this way, metaphorical thinking may yield new avenues of theorizing and empirical investigation about a phenomenon of interest. I will examine the heuristic and rhetorical uses of the machine metaphor in synthetic biology in the next two subsections.

### *The heuristic function of the machine metaphor in synthetic biology*

As already noted, there is general agreement that the heuristic use of metaphors in science is both ubiquitous and fruitful. It is fair to say that the machine metaphor has served life scientists well as a means of generating new ideas and hypotheses, and it is arguably the organizing metaphor of the whole discipline of synthetic biology. Thus, the authors of the articles on synthetic biology discussed all seem to recognize that thinking about living systems as if they are machines is extremely useful for the purpose of investigating and redesigning them.

Andrianantoandro et al. (2006, 12) emphasize that the usefulness of the machine metaphor is that it provides "a tool for conceptualization," which in the context of their discussion can reasonably be understood as support for the heuristic value of the machine metaphor. Thus, when synthetic biologists transfer engineering principles, methods, and concepts (such as standardization, abstraction, modularity, predictability, reliability, and uniformity) to biological systems, they should not "directly adopt" these principles, methods, and concepts (Andrianantoandro et al., 1). After all, there are also important differences between your ordinary clock, car, or computer and organisms, including autonomy, evolvability, and, perhaps more controversially, intrinsic purposiveness.[6]

Endy also seems to recognize the heuristic value of conceiving of organisms as engineered machines when he writes:

> Failures [to create an engineering technology based on living systems] would directly illuminate and help prioritize the most relevant gaps in our current understanding of natural living systems, and suggest how we might best eventually come to understand and apply nature's original technology.
>
> (Endy 2005, 449)

I suggest that synthetic biologists and the field as such can be understood as taking the heuristic use of the machine metaphor to its limits. It is, at least in part, the attempt to bring the heuristics of analyzing or decomposing living systems into modular parts with intrinsic specific functions into practice by attempting to build (or recompose) living systems that behave in a (more) linear and aggregative fashion.[7] However, as Endy notes, even failing to do so is another way of gaining knowledge about naturally evolved life and its organization.

Coming back to Black's interactionist theory, one of his claims was that metaphor is *bidirectional*. The idea is that when we say that "organisms are machines," this need not merely inform the way in which we come to think about organisms. It may as well shape the way we think about machines and engineering processes. One of the results of the heuristic application of the machine metaphor in synthetic biology has been an increasing emphasis among its practitioners on the way in which engineering processes and products in some domains can usefully be understood in terms of the processes and products of biological evolution. O'Malley (2011) draws attention to the widespread use of "kludging" in software engineering (see also Calcott et al. 2015). What these authors argue is that there is not a sharp contrast between the way engineering proceeds and the nature of engineered systems on the one hand and biological evolution and the nature of living systems on the other.

Summing up, in the context of synthetic biology, the heuristic use of the machine metaphor seems to work both ways and pave the way for a more nuanced understanding of the nature of engineering itself.[8] Still, it should be kept in mind that applying the machine metaphor heuristically does not entail that one also thinks that organisms and machines are of the same ontological kind. As several authors have highlighted (Nicholson 2013; Boudry and Pigliucci 2013), there is a genuine danger that the heuristic power of the machine metaphor might draw scientists into assuming that organisms basically are the same kind of thing as a washing machine or a computer. I have argued that synthetic biologists seem to be well aware of the important differences between a clock and a cockroach that make an ontological identification objectionable (Holm 2015). However, the heuristic use of a metaphor may come to inform nonscientific discourse and communication in problematic ways. I will consider this issue further in the next section.

### *The rhetorical function of the machine metaphor in synthetic biology*

Like most other life scientists, synthetic biologists are quick to appeal to the machine metaphor when they communicate their research to nonexperts. I will argue that a strong case can be made against the widespread and uncritical rhetorical use of the machine metaphor in particular and metaphors in general.

Nicholson criticizes the rhetorical use of the machine metaphor or what he calls "the machine conception of the organism" (MCO):

> The rhetorical use of the MCO by biologists [can] inadvertently mislead non-specialists into assuming that organisms really are machines.
>
> (2013, 675)

Nicholson points to some of the repercussions of the rhetorical use of the "molecular machines" metaphor introduced by Alberts (1998). After proposing that molecular biologists should view cells as factories composed of "assembly lines of protein machines," the metaphor has found its way into the literature, but mainly as a rhetorical device found almost exclusively in review articles; in the titles, abstracts, and introductions of research articles; and in titles of conferences. The term "molecular machine" and its associates also regularly crop up in science journalism, in interviews with scientists, and in popular illustrations of how cells work. Thus, the metaphor is channeled into the public discourse on molecular biology. However, Nicholson points out, the widespread rhetorical use of the machine metaphor has resulted in harmful consequences that "far outweigh its potential pedagogical and sociological benefits" (2013, 676; see also Pigliucci and Boudry 2011; Boudry and Pigliucci 2013 for similar claims).

The consequences that Nicholson has in mind concern the appropriation of the machine metaphor by contemporary creationists. Because scientists themselves speak in terms of machine metaphors, creationists are given the opportunity to clothe their claims about intelligent design in the vocabulary of mechanistic scientists.

Here's an example of what Nicholson has in mind. In an interview with *The Christian Post*, Michael Behe, a creationist and prominent proponent of intelligent design, is asked, "Do you see ID having enough evidence?" He replies:

> Yes, I certainly do. Well, I am a biochemist and biochemistry studies the molecular basis of life. And in the past 50 years, science has discovered that at the very foundation of life there are sophisticated molecular machines, which do the work in the cell. I mean, literally, there are real machines inside everybody's cells and this is what they are called by all biologists who work in the field, molecular machines. They're little trucks and busses that run around the cell that takes [*sic*] supplies from one end of the cell to the other. They're little traffic signals to regulate the flow. They're sign posts to tell them when they get to the right destination. They're little outboard motors that allow some cells to swim. If you look at the parts of these, they're remarkably like the machineries that we use in our everyday world.
>
> The argument is that we know from experience that machinery in our everyday world that we use in our everyday world required design, required an intelligent agent that put it together, who understood how it was going to be used and who assembled the parts. By an inductive argument, when we find such sophisticated machinery in other places too, we can conclude that it also requires design. So now that we found it in life and in the very foundation of life, I and other ID advocates argue that there is no reason to not reach the same conclusion and that in fact, these things were indeed designed.
>
> (Behe 2005)

Behe trades in Alberts's notion of molecular machines to boost the scientific credentials of ID. If molecular biologists themselves claim that cells are composed of little fine-tuned machines, then it seems like a genuine scientific hypothesis to propose that they are the product of divine design. In this way, creationists such as Behe can sidestep Humes's original rejection of the argument from design, which is based on the premise that while there are interesting similarities between, e.g., organisms and man-made machines, it is a mistake to think that organisms are, ontologically speaking, machines.

While Nicholson is right to point out that creationists such as Behe trade in the use of the machine metaphor, I think it is important to recognize that Behe could say of cells that "this is what they are called by all biologists who work in the field, molecular machines" even if scientists never used the machine metaphor outside the scientific community in which it, let's assume, serves a strictly heuristic function.

The rhetorical use of machine metaphors in synthetic biology has also been criticized for potentially confusing a range of questions concerning ethics and policy-making with regard to synthetic biology (Boldt 2018; see also Christiansen's article in this volume). Boldt claims that "it is ethically imperative to resist the thrust of machine metaphors" because such metaphors "hide" other important features of organisms such as "evolutionary change and interdependencies in ecosystems," which are relevant from the point of view of risk evaluation, e.g., in connection with the introduction of synthetic organisms into natural environments.

Questions concerning the moral status of living beings may, according to Boldt, also be influenced by the widespread appeal to machine metaphors. We do not normally think that man-made machines such as cars and computers deserve our direct moral consideration, so if scientists are taken to think that organisms are machines due to their rhetorical use of the metaphor, this may unjustifiably discredit bioethical arguments that rely on a rejection of the ontological identification of organisms and machines.

Finally, as is clear from the way in which synthetic biology is characterized by its main proponents, the field is to a large extent growing on the basis of its promise to be able to help solve a range of big problems facing our civilization with regard to health, food, and energy. By framing the communication of their research in terms of the machine metaphor, synthetic biologists are at risk of making decision makers assume that their solutions are more reliable and effective than they really are (given the true nature of living systems). It may also give synthetic biology an unfair advantage over alternative approaches when it comes to political decisions about how best to solve the problems in question.

What these criticisms of the rhetorical use of the machine metaphor have in common is that they highlight how this metaphor is likely to reinforce misconceptions about the fundamental nature of living systems – misconceptions that may lead to negative intellectual, ethical, and societal consequences. Moreover, there is a genuine risk that not only nonexperts but also scientists themselves will have their thinking about living systems overly shaped by machine thinking and thereby neglect the aspects of living systems that are suppressed, as Black would say, by the machine metaphor. In the next and final section, I consider just how strongly metaphors have been shown to influence the way humans think about a topic.

## Framed by metaphor: the influence of metaphor on reasoning

I have argued that while there might be great heuristic value to machine metaphors when they are used *within* a scientific community, using metaphors rhetorically and heuristically also presents important problems, such as the risks of misleading nonexperts and of being used to dress up nonscientific theories in scientific clothing. My discussion has mainly engaged with philosophers debating the proper function of metaphors in science and science communication. In this section, I will draw attention to empirical evidence that supports the view that scientists should be very careful when drawing on metaphors to convey understanding of scientific knowledge and practices.

Metaphorical framing has been experimentally shown to wield strong influence on human reasoning. In their investigation, Thibodeau and Boroditsky (2011) are concerned with the way in which metaphors may influence reasoning about crime. Their study shows that metaphors play a significant role in how people reason about crime. Importantly, in virtue of shaping how people think about crime, metaphors also indirectly exert a strong influence on policy-making concerning crime – i.e., different metaphorical descriptions of topics of social significance are likely to have very different practical consequences. (This is in line with Boldt's claim that the machine metaphor may have practical consequences due to its impact on the way laymen, policy makers, and regulators reason and assess proposed biotechnological solutions to societal problems.)

In their study, Thibodeau and Boroditsky "focus on two contrasting metaphors for crime: crime as a virus and crime as a beast." They test for whether these metaphors "subtly encourage people to reason about crime in a way that is consistent with the entailments of the metaphors." The hypothesis is that people's responses to treatment of crime will be influenced by the metaphorical framing of the problem:

> For example, might talking about crime as a virus lead people to propose treating the crime problem the same way as one would treat a literal virus epidemic? Might talking about crime as a beast lead people to propose dealing with a crime problem the same way as one would deal with a literal wild animal attack?
>
> (Thibodeau and Boroditsky 2011, e16782)

At the initial stage of their investigation, Thibodeau and Boroditsky established that, when asked about how to deal with the literal problem of a "virus infecting a city," respondents universally suggested solutions to the problem that focused on identifying and rooting out the source of the virus and focusing on preventing such an epidemic from occurring. When asked about how to deal with the literal problem of a "wild beast preying in a city," they universally suggested solutions focused on capturing and killing or caging it.

Having established such "schematic representations for solving literal virus or beast problems," the central question is whether those representations "transfer to people's reasoning about crime if crime is metaphorically framed as a virus or a beast." To see whether the metaphorical framing of crime problems plays a role in the reasoning and solutions people propose to crime problems, Thibodeau and Boroditsky performed five experiments. The conclusions they draw from those five experiments are as follows:

1    When crime was framed metaphorically as a virus, participants proposed investigating the root causes and treating the problem by enacting social reform to inoculate the community, with emphasis on eradicating poverty and improving education. When crime was framed metaphorically as a beast, participants proposed catching and jailing criminals and enacting harsher enforcement laws.

2    Single words are sufficient to instantiate the metaphorical frames that result in different proposals for dealing with the problem. Thus, participants "spontaneously extracted the relevant relational inferences even given a single metaphorical noun."

3    There is no evidence that "simply hearing a word like beast, even outside of the context of crime, would activate representations of hunting and caging." Thus, activation of "lexical associates" does not "bleed into people's descriptions of how to solve the crime problem." Importantly, this suggests that "metaphors act as more than just isolated words – their power appears to come from participating in elaborated knowledge structures."

4    The study also showed that metaphorical framing is likely to affect how people will seek out information for future problem-solving. Thus, when "given the opportunity to gather further information about the issue, participants chose to look at information that was consistent with the metaphorical frame."

5    It matters *when* the metaphorical framing takes place. Early metaphorical framing had a significant effect on people's suggestions for solving the problem as well as their information-seeking, whereas late metaphorical framing had no such effect. Thus, the study indicates that "metaphors can gain power by coercing further incoming information to fit with the relational structure suggested by the metaphor."

A general conclusion that Thibodeau and Boroditsky draw on the basis of their experiments is that the influence that metaphor exerts on our reasoning about a topic is covert. Thus, participants do not themselves identify metaphorical framing as relevant for their reasoning, but instead highlight something that is shared by both groups – namely, crime statistics – as a main influence.

The study seems to cohere well with the interactionist view that metaphors construct, control, and organize the way in which we think about the target. However, one might ask how strongly metaphor frames influence our thinking. To ascertain the strength of metaphors, Thibodeau and Boroditsky compared the influence of metaphorical framing with the predictable relationship that exists between people's political affiliation and gender, respectively. Both political affiliation and gender were shown to be statistically significant for predicting whether people would emphasize enforcement in response to crime. However, the study also showed that the two metaphorical frames produced even larger differences in opinion than the variance that corresponded to differences in gender and political affiliation.

Summing up, Thibodeau and Boroditsky's experimental study supports the view that metaphorical framing has a significant impact on how people understand, seek information, and reason about complex issues.

## Conclusion

The documented significance of metaphorical framing gives empirical support to some claims about the role of metaphors in science. Lewontin (2001) quotes Arturo Rosenblueth and Norbert Wiener as saying that "the price of metaphor is eternal vigilance." I think the study by Thibodeau and Boroditsky corroborates this point, and it sums up nicely the lessons of this chapter's discussion of the use of the machine metaphor in synthetic biology and elsewhere. When their whole field is organized in large part around the machine metaphor, there is a genuine risk that synthetic biologists will slide from heuristic uses to ontological assumptions about the nature of organisms. However, many synthetic biologists find that their research makes them painfully aware of the limits of the machine metaphor (see Holm 2015 for further discussion). The widespread application of the machine metaphor is, I think, more worrying when applied as a means for communicating scientific research and knowledge to nonspecialists.

Black's original analysis of metaphor provides an important insight: metaphors do not merely highlight similarities; they shape the way we think about the target, and, as shown by Thibodeau and Boroditsky, they may have significant impacts on our reasoning in ways that are not transparent to us. As soon as scientists introduce the machine metaphor to explain what they are doing, they frame people's thinking about biological systems. Saying that a cell consists of molecular machines or that genes are the blueprint of the organism does more than communicate a specific piece of knowledge about a specific biological entity. It might be said to have a "cascade effect" with regard to the way people come to think about the domain in question. Such effects might not be harmless. One effect might be that nonexperts form a range of false beliefs about the domain in question due to their unsupervised application of the metaphor. Another effect might be that practical decisions are guided on the basis of such metaphorically generated false beliefs or misunderstandings. In short, we have very good theoretical and practical reasons for encouraging scientists to handle the machine – as well as other – metaphors with care.

## Acknowledgments

I would like to thank Lucy Holt and Andreas T. Christiansen for their valuable comments on a draft of the chapter and the Independent Research Fund Denmark grant number DFF – 4180–00146 for funding my research for this article as part of the project Living Machines? Prospects and Limitations of the Machine Conception of the Organism.

## Notes

1  A similar view is proposed by Hesse (1966), who contends that theoretical explanations invoking new theoretical language are "metaphorical redescriptions" of the target domain. Other influential treatments of metaphor include Boyd (1993), Bradie (1998), Lakoff and Johnson (2003), and Gentner et al. (2001).
2  For excellent overviews of the heterogeneity and unifying features of synthetic biology research approaches, see O'Malley et al. (2008) and Deplazes (2009).
3  "Standardization" involves the definition, description, and characterization of the basic biological parts to be combined to produce functional biological devices and systems. "Decoupling" signifies the process of breaking down the construction of complex entities into independent and manageable tasks. Finally, "abstraction" concerns the identification of hierarchies of functional units and thus serves to enable the separation of design and fabrication (Endy 2005, 450).
4  This is a feature that enables reliable modelling and simulation, something that is also highly desirable for efficient design of biotechnological systems (Heinemann and Panke 2006, 2796).
5  This distinction is suggested in Bradie (1999), who adds a third, theoretical, function. The merits of the theoretical function of the machine metaphor are discussed in Nicholson (2013) and Holm (2015).
6  For an informative discussion of the differences between organisms and machines, see, e.g., Nicholson (2013).
7  My interpretation of Endy is inspired by Bechtel and Richardson's characterization of the heuristics of decomposition and localization (Bechtel and Richardson 1993).
8  I take Calcott (this volume) to make a strong case for a similar claim regarding the notion of a program.

## References

Alberts, B. (1998). The cell as a collection of protein machines: Preparing the next generation of molecular biologists. *Cell*, 92, 291–294.
Andrianantoandro, E., Basu, S., Karig, D. K., & Weiss, R. (2006). Synthetic biology: New engineering rules for an emerging discipline. *Molecular Systems Biology*, 2, 1–14.
Ball, P. (2004). Synthetic biology: Starting from scratch. *Nature*, 431, 624–626.
Bechtel, W., & Richardson, R. (1993). *Discovering Complexity: Decomposition and Localization as Strategies in Scientific Research*. Princeton: Princeton University Press.
Behe, M. (2005, May 27). Understanding creation, evolution and intelligent design. *The Christian Post*. Accessed February 14, 2019.
Black, M. (1955). Metaphor. *Proceedings of the Aristotelian Society*, 55, 273–294.
Black, M. (1993). More about metaphor. In A. Ortony (ed.), *Metaphor and Thought*. Cambridge: Cambridge University Press.

Boudry, M., & Pigliucci, M. (2013). The mismeasure of machine: Synthetic biology and the trouble with engineering metaphors. *Studies in History and Philosophy of Biological and Biomedical Sciences*, 44, 669–678.

Boyd, R. (1993). Metaphor and theory change: What is 'metaphor' a metaphor for? In A. Ortony (ed.), *Metaphor and Thought*, 2nd ed., 481–532. Cambridge: Cambridge University Press.

Bradie, M. (1998). Explanation as metaphorical redescription. *Metaphor and Symbol*, 13, 125–139.

Bradie, M. (1999). Science and metaphor. *Biology and Philosophy*, 14, 159–166.

Calcott, B., Arnon, L., Siegal, M. L., Soyer, O. S., & Wagner, A. (2015). Engineering and biology: Counsel for a continued relationship. *Biological Theory*, 10(1), 50–59.

Deplazes, A. (2009). Piecing together a puzzle. *EMBO Reports*, 10, 428–432.

Endy, D. (2005). Foundations for engineering biology. *Nature*, 438, 449–453.

Endy, D. (2008). Engineering biology–A talk with Drew Endy. Interview in online magazine *Edge*. Accessed January 15, 2015.

Gentner, D., Bowdle, B., Wolff, P., & Boronat, C. (2001). Metaphor is like analogy. In D. Centner, K. J. Holyoak, & B. N. Kokinov (eds.), *The Analogical Mind: Perspectives from Cognitive Science*, 199–253. Cambridge, MA: MIT Press.

Heinemann, M., & Panke, S. (2006). Synthetic biology–putting engineering into biology. *Bioinformatics*, 22, 2790–2799.

Hesse, M. (1966). *Models and Analogies in Science*. Notre Dame, IN: University of Notre Dame Press.

Holm, S. (2015). Is synthetic biology mechanical biology? *History and Philosophy of the Life Sciences*, 37(4), 413–429.

Keasling, J. (2005). The promise of synthetic biology. *Bridge National Academy of Engineering*, 35, 18–21.

Lakoff, G., & Johnson, M. (2003). *Metaphors we Live By: With a New Afterword*. Chicago: University of Chicago Press.

Lewontin, R. C. (2001). In the beginning was the word. *Science*, 291, 1263–1264.

Nicholson, D. (2013). Organisms machines. *Studies in History and Philosophy of Biological and Biomedical Sciences*, 44, 669–678.

O'Malley, M. (2011). Exploration, iterativity and kludging in synthetic biology. *Comptes Rendus Chimie*, 14, 406–412.

O'Malley, M., Powell, A., Davies, J. F., & Calvert, J. (2008). Knowledge-making distinctions in synthetic biology. *BioEssays*, 30, 57–65.

Pigliucci, M., & Boudry, M. (2011). Why machine information metaphors are bad for science and science education. *Science & Education*, 20, 453–471.

Reynolds, A. S. (2018). *The Third Lens: Metaphor and the Creation of Modern Cell Biology*. Chicago: University of Chicago Press.

Thibodeau, P. H., & Boroditsky, L. (2011). Metaphors we think with: The role of metaphor in reasoning. *PLoS One*, 6(2), e16782.

Yokobayashi, Y., Weiss, R., & Arnold, F. H. (2002). Directed evolution of a genetic circuit. *Proceedings of the National Academy of Sciences of the United States of America*, 99(26), 16587–16591.

# Editors' postscript

Having finished the work on this volume, we would like to highlight some of the themes we think stand out as well as some of the new kinds of questions and future research avenues we think the volume points to.

One theme concerns the kinds of explanations that work or are adequate in interdisciplinary fields like systems biology and synthetic biology. While questions about the differences and compatibilities among these kinds of biological explanations already have a tradition in philosophical debates, the issue of how confirmation and explanation are related in these fields has received less attention. We suggest that one way to make progress on these problems is to pay more attention to the relata of confirmation and explanation relations: phenomena and "theoretical" representations.

Another important set of issues raised in this volume (via the examination of recent methodological developments in systems and synthetic biology) concerns the challenges for the integration of multiple methods, accounts, and data required for producing adequate explanations of complex biological phenomena. Reflecting on the limitations of explanatory integration in systems and synthetic biology raises the wider issue of the conditions under which intelligible accounts of contemporary science, more generally, are possible. Thus, one of the primary benefits of studying problems of explanation in interdisciplinary fields like systems biology is that it puts in context questions about methodology in science as well as more general epistemological questions concerning meaning and the limits of intelligibility.

Several contributions in this volume highlighted the fact that systems biology and synthetic biology are premised on interdisciplinary collaboration of mathematical modelling and laboratory experimentation as well as on the application to biology of concepts and tools from physics, computer science, and engineering. These practices raise general philosophical questions (e.g., How should concepts and methods from different fields be combined? What should we understand the value of integrative systems biology to be?) as well as more practical or methodological questions (e.g., How to manage these interfield relationships? How to move beyond the manifesto rhetoric of a new field to the critical outlook necessary for healthy methodological practices?). The ambition (and hope) shared by many of this volume's contributors is for philosophy to play a role in identifying and resolving some of the problems facing the fields of systems biology and synthetic biology.

Finally, we think that several contributions to the volume document the historical and current significance of machine metaphors in the life sciences and the many dimensions along which the use of machine metaphors deserves careful study. The recognition that the price of metaphors requires eternal vigilance has both a theoretical and a practical aspect, and while several of the contributions to this volume discuss the ways in which machine metaphors may mislead, we also find that there is much more work to be done analyzing the ways in which scientists use metaphors and which challenges these uses raise both within scientific practice and in relation to the role of science in society.

The volume also suggests several new kinds of questions that one might ask. One general kind of question, which emerges from reflection on the practice of interdisciplinary or transdisciplinary fields like systems and synthetic biology, concerns the very *identity of science*. What are the primary defining directions along which individual research projects can be situated and their quality assessed? What is the distinction between basic science and engineering? Is there a general shift currently underway from scientific research to engineering? Addressing this sort of question(s) requires investigating in more depth the relation between science and society and its different subsystems. It also requires reflection on and conceptual investigation of what it means to orient research practices primarily toward understanding or construction and creation. What are the consequences of promoting a pluralism of orientations in hybrid understanding/engineering activities? Do these consequences affect the relation between science and society? Or the self-image of science?

Another set of questions concerns the practical and cognitive organization of science: how is it possible to perform science under current organizational and practical conditions (project-based research, technology and data-driven research, collaborative networks, multidisciplinary and multicenter approaches, etc.), and what kind of science results from it? Even if science is clearly and primarily oriented toward gaining new explanations and understanding, how do we conceive of this process in a Big Science context? How does this bear on the epistemic status of the individual mind and the individual scientist, which feature so centrally in traditional epistemological thought?

As science becomes increasingly more multidisciplinary, multisited, and collaborative, and focuses on scientific as well as technological and societal issues, (successful) communication has become a more central and challenging component of doing (good) science. What are the strategies currently used to secure terminological accuracy? What should the standards be for evaluating the soundness of the goals and visions of new disciplines/research directions?

While the chapters in the volume zoom in on particular research questions, we also think the volume as a whole suggests some more general avenues of research. One very broad avenue concerns a reevaluation and a reframing of what science is, how knowledge is produced, and what it means to know something, especially in the context of thinking about (techno)science or the relation between science and engineering (with their associated goals and visions). When it comes to the theme concerning the use of metaphors in science and its impact on, e.g., bioethics and science communication, we find that the volume points to the importance of a more in-depth exploration of the cognitive and social dimensions of science. Relatedly, the volume opens up an investigation into the role(s) that philosophy (and philosophers as public intellectuals) can play in managing the communication of both contemporary scientific research and the historical tradition of doing science.

# Index

Note: Page numbers in *italic* indicate a figure and page numbers in **bold** indicate a table.